International Biolaw and Shared Ethical Principles

T0252703

The Universal Declaration on Bioethics and Human Rights marked a significant step towards the recognition of universal standards in the area of science and medicine. More than a decade later, this book provides an overview of the ethical and legal developments which have since occurred in the field of bioethics and human rights. The work critically analyzes the Declaration from an ethical and legal perspective, commenting on its implementation, and discussing the role of non-binding norms in international bioethics. The authors examine whether the Declaration has contributed to the understanding of universal or global bioethics, and to what degree the states have implemented the principles in their domestic legislation. The volume explores the currency of the Declaration vis-à-vis the more recent developments in technology and medicine and looks ahead to envisage the major bioethical challenges of the next twenty years. In this context, the book offers a comprehensive ethical and legal study of the Declaration with an in-depth analysis of the meaning of the provisions, in order to clarify the extension of human rights in the domain of medicine and the obligations incumbent upon UNESCO member states, with reference to their implementation practice.

Cinzia Caporale is Head of the Research Ethics, Bioethics, Biolaw and Biopolitics research unit at the Institute of Biomedical Technologies of the National Research Council of Italy (CNR). She is the Coordinator of the CNR Research Ethics and Bioethics Committee and represents CNR on the topic of Research Integrity on the international level. She is Adjunct Professor of Bioethics at Sapienza University of Rome and a member of the Italian National Bioethics Committee. She is President of the Ethics Committee of L. Spallanzani National Institute for Infectious Diseases and of the U. Veronesi Foundation and Director of the scientific journal *The Future of Science and Ethics*. She has been Chairperson of UNESCO's Intergovernmental Bioethics Committee for two mandates. Under her presidency the Universal Declaration on Bioethics and Human Rights was elaborated and adopted.

Ilja Richard Pavone gained his PhD in International Law and Human Rights at Sapienza University, Rome. He is Researcher of International Law at the National Research Council (CNR) of Italy, Rome, where he coordinates the research unit in International, European and National Biolaw. He is also Professor of Environmental Law at Tuscia University, Viterbo. In the past, he has taught at the universities of Rome (Sapienza), Venice (Venice International University), Macerata and Siena. He has been Visiting Professor at Queensland University, New York University, Bochum University and Max Planck Institute for Comparative Public Law and International Law (Heidelberg). He is author of over fifty publications, essays and articles in International Law and European Union Law, with a particular focus on Bioethics, Human Rights, Animal Law, International Health Law and Environmental Protection.

International Biolaw and Shared Ethical Principles

The Universal Declaration on Bioethics and Human Rights

Edited by Cinzia Caporale and Ilja Richard Pavone

LONDON AND NEW YORK

First published 2018 by Routledge

2 Park Square, Milton Park, Abingdon, Oxfordshire OX14 4RN
52 Vanderbilt Avenue, New York, NY 10017

Routledge is an imprint of the Taylor & Francis Group, an informa business

First issued in paperback 2019

British Library Cataloguing-in-Publication Data
A catalogue record for this book is available from the British Library

Library of Congress Cataloging-in-Publication Data
Names: Caporale, Cinzia, editor. | Pavone, Ilja Richard, editor.
Title: International biolaw and shared ethical principles : the Universal Declaration on Human Rights and Bioethics / edited by Cinzia Caporale, Ilja Richard Pavone.
Description: New York, NY : Routledge, 2017. | Includes bibliographical references.
Identifiers: LCCN 2017007942 | ISBN 9781472483980 (hardback) | ISBN 9781317114406 (adobe reader) | ISBN 9781317114390 (epub) | ISBN 9781317114383 (mobipocket)
Subjects: LCSH: Biotechnology industries—Law and legislation. | Bioethics. | Biotechnology—Moral and ethical aspects. | Human rights.
Classification: LCC K3925.B56 I58 2017 | DDC 174.2—dc23
LC record available at https://lccn.loc.gov/2017007942

ISBN: 978-1-4724-8398-0 (hbk)
ISBN: 978-0-367-88209-9 (pbk)

Typeset in Galliard
by Apex CoVantage, LLC

Contents

Acknowledgements

The authors wish to express their appreciation for the excellent editorial support and valuable technical guidance provided by Giorgia Adamo during the editing process. The authors also extend their acknowledgements to Francis Allen Farrelly for the English linguistic review and to Ruth Noble for her patience, backing and encouragement.

Contributing authors

Francesco Alicino
Department of Political Science
University of LUISS 'Guido Carli', Italy
and
Department of Law
University of LUM 'Jean Monnet', Italy

Roberto Andorno
Institute of Law
University of Zurich, Switzerland

Carmela Decaro Bonella
Department of Political Science
University of LUISS 'Guido Carli', Italy
and
Department of Law
University of LUM 'Jean Monnet', Italy

Cinzia Caporale
Institute of Biomedical Technologies – ITB
National Research Council, Italy

Emilia D'Antuono
Department of Social Sciences,
Università degli Studi di Napoli, Federico II, Italy

Thomas Alured Faunce
Medical School and ANU College of Law
Australian National University, Australia

Adèle Langlois
School of Social and Political Sciences
University of Lincoln, United Kingdom

Fabio Macioce
Department of Law
Libera Università Maria Ss. Assunta LUMSA, Italy

Laura Palazzani
  Department of Law
  Libera Università Maria Ss. Assunta LUMSA, Italy

Ilja Richard Pavone
  Institute of Biomedical Technologies – ITB, Biolaw Unit
  National Research Council, Italy

Michèle Stanton-Jean
  Centre de recherche en droit public, Faculty of Law
  University of Montreal, Canada

Emilia Taglialatela
  Department of Social Sciences,
  Università degli Studi di Napoli, Federico II, Italy

Henk ten Have
  Center for Healthcare Ethics
  Duquesne University Pittsburgh, USA

Preface

Cinzia Caporale

From the theoretical point of view, universal declarations are rooted in two main ideas; the first is that each and every human being has the same value, irrespective of historical and geographical contingencies. The principles of liberty, dignity, equality and others belong to all human beings as such. Variety in morals, traditions, social and political structures, should not represent a cause or a reason for establishing differences in value among human beings. The second idea is the basic unity of all humankind. People who are separated by historical ages as well as by thousands of miles are all part of a same great community. They share the same origins and the same destiny. The more that different parts of the world become progressively interrelated, the more the ideal of the unity of humankind is supported by hard facts.

Universal declarations are then rooted in facts, but they are not just statements of facts. They aim at orienting collective and individual choices. In this way, universal declarations play a key role towards a progressive convergence of the several morals and legislations of the world.

However, the fact that the very idea of universal declarations derives from the assumption that all people have the same value and from the assumption of the unity of all humankind, should not be seen as an argument in favour of a progressive equalization of morals, cultures and even of political institutions in the world. The Unesco Declaration on Cultural Diversity, adopted in 2001, made clear to all that the universality of human rights was not tantamount to cancel or to weaken cultural differences. As article 1 states,

> Culture takes diverse forms across time and space. This diversity is embodied in the uniqueness and plurality of the identities of the groups and societies making up humankind. As a source of exchange, innovation and creativity, cultural diversity is as necessary for humankind as biodiversity is for nature. In this sense, it is the common heritage of humanity and should be recognized and affirmed for the benefit of present and future generations.

These premises and method were well kept in mind by the women and men who shared the challenging aim of elaborating the Universal Declaration on Bioethics and Human Rights (UDBHR) which, indeed, enucleated a set of values as general in scope as the values stated in the previous declarations and, at the same time, fully respected the plurality of morals and practices.

Bioethical principles and practices represent one of the deepest aspects of human identity and human life. Bioethics is the land of diversities and it must be considered as a plural noun. When it comes to bioethical identity, it is not clear which cultural and moral elements could be included, how they are to be interpreted or what their respective weight are. There is a lot of disagreement and passionate debate about this. Definitely, it can be stated that there is a tension between the supposition that there is such a thing as common principles and the fact that these common principles are neither well defined nor undisputed.

The cultural pluralism includes a plurality of moral convictions and ethical approaches. There is no monolithic morality and there never has been one. In cultural as well as in moral matters, we are and have always been a unity-in-diversity. Thus, when we argue about how to handle body, birth and death in the technological era of biomedicine along with cutting-edge research fields such as biotechnologies; synthetic biology; neurosciences and especially neurogenetics; the new technological wave of converging technologies and learning machines; and, in general, scientific innovation and its extraordinary impact on human life, society and the environment, we should take diversities very seriously. Main factors of divergence are religious, historical and cultural multiplicity of backgrounds, different philosophical traditions and legal systems and different mentalities. Furthermore, bioethics brings together two areas, science and ethics, that struggle with the tension of the universal and the particular, the global and the local. On a closer examination, it is also true that there are different understandings of the field itself, and different expectations about what the field can accomplish.

Bioethics not only shows differences about method and content, but also very important differences about how the field is defined and tensions between specific judgments and universal norms.

However, being a moral unity-in-diversity, the question cannot be whether differences in morally relevant areas of science and technology regulation are admissible. It can only be how much difference is admissible. Otherwise, maybe even better, how much unity, how much homogeneity is actually necessary?

The aim of the UDBHR elaboration process was thus to reach general consensus not by excluding or removing, but by including diversities. Furthermore, in order to find the final agreement, we were not guided by the aim of identifying principles that were so abstract as to become of little use as a practical guide in real world. Rather, we tried and possibly succeeded in identifying principles that include the different ethical views about human life and scientific progress that are considered as generally valid across the several cultures of the world.

In pursuing this goal, we kept in mind especially one of the two roots of universal declarations, namely, the basic unity of all humankind. As far as bioethics is concerned, in fact, this unity does not simply refer to the moral dimension of the fundamental questions of human life. As a matter of fact, nowadays peoples and countries of the world are also more and more interrelated. Medical knowledge, research and practices are increasingly shared at world level. Travels and migrations make different populations interact, and from this interaction new concerns and opportunities arise. As a consequence, in "issues related to medicine, life sciences and associated technologies as applied to human beings, taking into account their social, legal and environmental dimensions", to quote article 1 of the UDBHR, it is more and more difficult to separate actions that solely affect a country's internal affairs from those that have a significant impact on the internal affairs of other countries.

Increasingly, countries are faced with the fact that, in the fields of the global commons of bioethics and technological innovation, their legitimate powers have to be exercised in an interdependent way. Obviously, national states continue to perform crucial functions, and must have the powers to fulfill these functions effectively. However, these powers should be fully informed by the fundamental interests of humanity, in an inclusive rather than exclusive way. Moreover, the ever-increasing interrelation of our world should be seen as a great chance to pursue the values of individual and collective choices in medical treatment, easier access to good-quality health care for the greatest number of people, shared scientific progress amongst all countries, food security and improved nutrition, sustainable development, clean environment and clean and affordable energy et cetera.

In other words, the main objective of the paramount effort we all have made in elaborating and adopting the UDBHR was that of translating in the bioethical realm the universal values of liberty, equality, responsibility, social justice, pluralism, pursuit of knowledge.

The challenge was intellectually exciting and morally tricky. Feasible and meaningful.

And the potential not only to integrate but also to go beyond the existing international human rights instruments and international guidelines on bioethics was put into effect.

The reason for which the UDBHR was successfully adopted likely relies also on the fact that UNESCO, the ethics committees involved and every single Member State took seriously not only bioethics but likewise the idea itself of a real and sincere debate, because behind the idea of confronting opinions, a strong, demanding and provocative philosophical option is hidden. There is an implication which is that of a relational plurality. I am referring to that particular experience where relationships gather fundamental prominence: what we call "dialogue". The issue acquires particular importance when we do not approach mere abstract concepts, but when we talk about life stances, values, views about life and the world: exactly what bioethics deals with. The discussion on the UDBHR was positively dialogic. Dialogue is profoundly human and can take place only between people who recognize to each other the same value. Dialogue consists in the pursuit of agreement and consensus. It is then an inner attitude and it tangibly informed the elaboration of the Universal Declaration on Bioethics and Human Rights.

Practically, this meant that we had not simply to affirm the diversity of our bioethics identities and justify and legitimate what was already existing. On the contrary, we had to seek through dialogic criticism of the existing positions the deeper significance of bioethical problems and achieve a minimum commune, a sound shared bioethical vision.

Was it reasonable to imagine a global bioethics that could be universally recognized as legitimate and therefore adopted by everyone? Is it the request of the modern world for a shared vision of what is perceived as good – particularly when such presumption is applied to the frontiers of scientific experimentation or concerning people's rights over matters related to their body and applicable technologies – approachable or just illusory? Is it desirable?

Seldom there has been such a vigorous increase of interest for a new arena of confrontation in the last twenty years comparable to what has happened to bioethics, a popularity to which the UDBHR added value. Regarding innovation, bioethics has become a sort of common language, the area of equal rights that brings together the people and the scientists, users of technologies and the researchers who produced them. With two other positive effects: the first is that the appeal of the subject determines a deeper involvement of the citizens, minimizing the attitude to delegate decisions to 'experts' in that sort of subtle technocracy which regrettably affects this sector; the second concerns the recognition that bioethics has the merit to pose the question of an adequate information system based on evidence and an appropriate scientific description of the issues that must be tackled.

The year 2015 celebrated the decennial since the adoption of the UDBHR, a milestone in the process towards the construction of a global bioethics, a blueprint agreed to by almost all the world's countries in the golden age of UNESCO bioethics.

Are we prepared for the new serious and global ethical challenges we are facing? Does the Declaration maintain the forcefulness it had twelve years ago and, above all, what did we learn in terms of methodology and procedures to achieve an ethical minimum commune that can be applied for the future?

The collection of essays in this volume, all written by eminent academics, offers an original and insightful analysis of the modernity of the Declaration. All articles are written from a twofold perspective. They offer an in-deep evaluation of the main achievements of the discussion on global bioethics and international biolaw. And they are bold in prospecting the future developments of global bioethics and international biolaw in relationship to some of the most relevant issues for humankind, such as global justice and environmental sustainability.

The authors come from different cultural contexts, and are specialists in different fields. As they address a core of common issues, this ensures that the volume offers true interdisciplinary treatment and perspective.

I'm very grateful to them and particularly to Ilja Richard Pavone, whose research and academic career I had the privilege to support along the years.

Introduction

Cinzia Caporale and Ilja Richard Pavone

The adoption of the UNESCO Universal Declaration on Bioethics and Human Rights (UDBHR) in 2005 marked a significant step towards the recognition of universal standards in the fields of science and medicine. The UDBHR is the first international instrument devoted to bioethics within a human rights framework, whose ultimate goal is that of establishing internationally accepted principles for the construction of an 'international biolaw'. Its scope is wide, adding issues like life sciences, the protection of the environment, the biosphere and biodiversity, to the traditional field of medicine. The Declaration also shifted its focus from the individual to the society and humankind in general, extending its scope beyond classical bioethical topics such as individual autonomy, informed consent and patient–physician relation. Indeed, the Declaration incorporates new principles drawn from environmental ethics, such as the need to share knowledge and benefits derived by the advancement of science, the precautionary approach, the responsibility towards future generations and the protection of biodiversity as a common concern of humankind.

This volume discusses the role of the UDBHR in the route towards the development of a global bioethics and the implications of framing bioethical issues in the international legal discourse. Moreover, the volume opens with the observation that soft law approaches to questions of bioethical relevance have become the dominant framework for addressing many ethical questions at national and international level. In particular, three important UNESCO declarations (Human Genome, Genetic Data, and Bioethics and Human Rights) have been implemented so far by various States in this manner.

It is in this context of strong connections between bioethics, human rights, international law, international legal discourse and need of compliance, that this volume seeks to analyze what exactly happens when issues of high bioethical relevance are discussed and negotiated within an international framework.

After more than ten years from the adoption of the UDBHR, the time has come to enrich the bioethical debate with an in-depth scholarly analysis of the concrete impact of the Declaration on the construction of shared and common ethical principles. This is precisely the purpose of the present collection of essays, whose publication offers an opportunity not only to celebrate the tenth anniversary of the Declaration, but also to create new momentum towards its critical assessment and to promote its sound knowledge.

International Biolaw and Shared Ethical Principles gathers the essays of some of the world's leading scholars in the field of bioethics and biolaw to explore key ethical and legal issues which have emerged from the adoption of the Declaration. Specifically, we ask whether the Declaration is still relevant to the challenges raised by the continuous development of life science, or whether it should be updated, by including, for instance, issues related to neuroethics, animal ethics, biotech.

Has the Declaration solved the dilemma of how to conciliate multiculturalism and the need of shared principles in the domain of bioethics? Has it contributed to the understanding of universal or global bioethics? Have States implemented (and to what degree) the principles set out in the Declaration? What will be the major bioethical challenges over the next twenty years? What is the role of non-binding norms in international bioethics?

The volume draws on the strong expertise of the authors, addressing the aforementioned issues that are common to ethical theory, legal philosophy and international law. It is divided into two parts. Part I ('UNESCO and Global Bioethics') analyzes the main bioethical aspects surrounding the Declaration, with particular reference to the problems of the relationship between bioethics and human rights, and of how to conciliate different ethical views in the framework of the identification of common shared principles on these sensitive topics.

The volume begins with a reconstruction of the role of UNESCO in promoting universal human rights, from its foundation to the adoption of the UDBHR, claiming that the Declaration is an extension of human rights law, represented by the Universal Declaration on Human Rights, to the domain of biomedicine (Chapter I). It continues with a deep analysis of the negotiation process which led to the adoption of the Declaration (Chapter II), determining if and to what extent, the UDBHR represents a landmark in the development of a global bioethics (Chapter III). Subsequently, the never-ending dilemma of how to reconcile preserving multiculturalism when attempting to impose/identify common principles (Chapter IV), and the philosophical thread of the Declaration (Chapter V), are dealt with. Specifically, these chapters address the question of whether the Declaration provides a proper and satisfactory answer to critics of cultural relativism, on account of being based on principles – such as autonomy, social responsibility, solidarity, personal wellbeing – that belong to the Western approach to bioethics.

Part II ('The Contribution of the Universal Declaration on Bioethics and Human Rights to the Construction of a Shared Bioethics') is devoted to the role of UNESCO in setting standards in the field of bioethics and human rights. Part II begins with an analysis of the regulation of transnational practices in the Declaration focusing on the bioethical aspects related to clinical trials carried out in developing countries (Chapter VI).

The form and content of the Declaration (Chapter VII) and the role of soft law in bioethics (Chapter VIII) are then analyzed. These two Chapters deal with the peculiarity of soft law in promoting legislative responses to emerging challenges posed by biomedical advances; and second, they illustrate to which extent soft law – highly used in environment and outer space as a basis for the development of customary norms or for the future negotiation of binding treaties – can be a suitable tool for legislating in the emerging field of international biolaw. Later Part II will explore whether the UDBHR provisions, such as the social responsibility principle, technology transfer aspects and the focus on future generations, are likely to facilitate the normative transition from an epoch characterized by the political and economic power of multinational corporations (the Corporatocene), to an epoch characterized by the use of human technology to indefinitely sustain not only its own species but all lifeforms on the planet (the Sustainocene) (Chapter IX). The final chapter is devoted to the 'future developments', focusing on the role of UNESCO's International Bioethics Committee in promoting the principles of the UDBHR (Chapter X).

From a bioethical point of view, the volume aims at underlining the invaluable role of UNESCO, and in general, of global intergovernmental forums, in setting the foundations of plausible ethical principles and procedures which take into account different perceptions of ethical and legal issues and allow a fuller transnational scientific and biomedical cooperation. The UDBHR played a key role towards progressive convergence of the several morals and bio-legislations present in the world, although it did not just favor a mere ethical equalization, but it contributed to a shared definition of bioethics, an achievement which could not been imagined at the very beginning of the crafting process.

From a legal point of view, the volume aims firstly at proving the potential of non-binding instruments in the promotion of adequate legal answers to the challenges posed by medicine and science, without holding back on the intrinsic limits of soft law. It attempts to model the normative presuppositions of existing theories on the role of soft law, in terms that take into account developments in the theoretical debate which has occurred over the past two decades, especially in environment protection and outer space governance. The main aim is, therefore, that of giving conceptual instruments for translating the background on the role of non-binding norms in environment protection and outer space governance in the wide sector of science and innovation and in particular of biomedicine and life sciences. Secondly, the volume evaluates whether UNESCO declarations in the field of bioethics can be considered as part of an emerging 'international biolaw' and, finally, it analyzes to which extent the Declaration has the potential to evolve, like the Universal Declaration on Human Rights, into international customary law or be the basis for an international biolaw convention.

The conclusions will situate the UDBHR in the broad discourse on the evolution of international law in general, from 'international law of coexistence' towards the 'international law of cooperation', to shed new light on some of the specificities and shortcomings of UNESCO standard-setting pointed out in the text.

On a final note, this work will investigate the possibility (the ambition) to achieve an ideally 'shared bioethics' which could then feed and enhance the human rights framework. In this context, it will evaluate the potentiality of soft law, which must be considered by States as the most effective legal tool to regulate within their domestic legislations the controversial bioethical issues related to the advancement of life sciences.

The editors sincerely hope that this volume will serve as a vehicle to improve knowledge of the Declaration, by favoring the process of strengthening the awareness of existing human rights in the field of life sciences, in the perspective of promoting a global bioethics addressed to the most vulnerable population groups.

Part I

UNESCO and global bioethics

I   The role of UNESCO in promoting universal human rights

From 1948 to 2005

Roberto Andorno

Introduction

UNESCO was created in the aftermath of the Second World War to reaffirm the conviction of the international community that intercultural dialogue and respect for justice and human rights are essential to build a durable peace. The UNESCO Constitution, which was adopted in November 1945, states that the first objective of the organization is "to contribute to peace and security by promoting collaboration among the nations through education, science and culture in order to further universal respect for justice, for the rule of law and for human rights and fundamental freedoms".

Consistent with this goal, UNESCO formed in 1947 a committee of intellectuals from different countries and cultural backgrounds, who made recommendations for the development of a universal human rights instrument. In this way, the Organization contributed to the preparatory work of the *Universal Declaration of Human Rights* (henceforth UDHR) of 1948, which would become the pillar upon which the entire human rights system is built. Almost 60 years later, in 2005, UNESCO served as a platform for the international community to develop the *Universal Declaration of Bioethics and Human Rights* (henceforth UDBHR), which is the first global legal instrument that comprehensively addresses the linkage between human rights and bioethics.

This chapter aims to draw a parallel between these two significant efforts of UNESCO in the promotion of universal human rights, and to emphasize that the UDBHR is ultimately an extension of international human rights law to the specific field of biomedicine.

A short history of UNESCO

Towards the end of 1942, in the middle of the Second World War, representatives of the European countries that were fighting Nazi Germany and its allies had a first meeting in London for what came to be known as the Conference of Allied Ministers of Education (CAME). They were examining ways and means to reconstruct their systems of education once the war had come to an end. In one of the subsequent meetings of CAME, it was proposed to create an international organization for education. Upon this proposal, the recently founded United Nations convened a conference in London in November 1945 for "the establishment of a United Nations Educational and Cultural Organization".[1]

---

1   F. Valderrama, *A History of UNESCO*, UNESCO, Paris 1995, pp. 19–24.

Representatives of 44 countries took part at the London Conference, which was presided over by Ellen Wilkinson, Minister of Education of Great Britain. The delegates decided to create an organization that would embody a genuine culture of peace and establish the "intellectual and moral solidarity" of humankind in order to prevent the outbreak of another world war.[2] The driving idea shared by delegates was that fostering education and intercultural dialogue was the best way to promote peaceful relationships between countries. This inspiring idea was thereafter enshrined at the beginning of UNESCO's Constitution, which states that "since wars begin in the minds of men, it is in the minds of men that the defences of peace must be constructed".

It is noteworthy that the initially planned organization was only concerned with educational and cultural issues. However, a number of scientists were pressing for the inclusion of science in both the title of the organization and in its programme of activities. In the forefront of this effort to put the "S" for science in the organization's name and goals were the British biologist and philosopher Julian Huxley, who would then become the first General Director of UNESCO, and the British biochemist Joseph Needham.[3] Decisive support for the inclusion of science within the scope of the planned organization was the recent dropping of the atomic bomb on Hiroshima and Nagasaki. This dramatic event would brutally show the tremendous ambivalence of science and technology, which can be used for the best and for the worst. The mushroom clouds over Japan had suddenly made the ethics of scientific research the burning question of the day. It is therefore not surprising that in her opening speech at the Conference, Wilkinson declared:

> Though Science was not included in the original title of the Organization, the British delegation will put forward a proposal that it be included, so that the title would run "Educational, Scientific and Cultural Organization". In these days, when we are all wondering, perhaps apprehensively, what the scientists will do to us next, it is important that they should be linked closely with the humanities and should feel that they have a responsibility to mankind for the result of their labours. I do not believe any scientists will have survived the world catastrophe, who will still say that they are utterly uninterested in the social implications of their discoveries.

These words of Ellen Wilkinson summed up the anxieties felt by a majority of the delegates. On 6 November 1945, the "S" for Science was finally incorporated into the title of the new body, which should play a "humanization role in the education of scientists".[4] At the end of the conference, 37 countries founded the new UN agency. The Constitution of UNESCO, signed on 16 November 1945, came into force on 4 November 1946 after ratification by 20 countries: Australia, Brazil, Canada, China, Czechoslovakia, Denmark, Dominican Republic, Egypt, France, Greece, India, Lebanon, Mexico, New Zealand, Norway, Saudi Arabia, South Africa, Turkey, the United Kingdom and the United States. The first General Conference of the Organization was held in Paris from 19 November to 10 December 1946 with the participation of representatives from 30 member states entitled to vote and from 18 non-member states.[5]

---

2 UNESCO Constitution (1945), Preamble.
3 F. Valderrama, *op. cit.*
4 A. Plomer, *Patents, Human Rights and Access to Science*, Edward Elgar Publishing, Cheltenham 2015, p. 141.
5 F. Valderrama, *op. cit.*

Today, after 70 years of existence, UNESCO comprises 195 member states, that is, virtually all states in the world. In addition to its headquarters in Paris, the Organization has more than 50 field offices around the world. UNESCO implements its activities through the five program areas of Education, Natural Sciences, Social and Human Sciences, Culture, and Communication and Information. Sometimes referred to as "the intellectual agency" of the United Nations, UNESCO describes itself as a "laboratory of ideas and a standard-setter to forge universal agreements on emerging ethical issues".[6] Today, maybe more than ever, the Organization plays a crucial role in fostering intercultural dialogue and, at the same time, promoting respect for human rights.

UNESCO and human rights

From the very beginning of its foundation, UNESCO was inextricably linked to the human rights movement that emerged in the aftermath of the Second World War. In December 1948, the General Assembly of the recently created United Nations adopted the Universal Declaration of Human Rights. While this document was still being drafted by the Human Rights Commission, UNESCO decided to anticipate the philosophical questions that the elaboration of such a declaration would inevitably raise: Could any values be said to be common to all countries? What would it mean to speak of certain rights as "universal"?

In order to address these challenging questions, UNESCO recruited some leading thinkers of the time for a "Committee on the Theoretical Bases of Human Rights". This panel, chaired by the British historian and diplomat Edward H. Carr, prepared a questionnaire dealing with various theoretical problems in the formulation of an "international declaration of human rights", and sent it out to scholars and statesmen around the world. Among the notable figures who responded to the 8-page questionnaire were Mahatma Gandhi, Jacques Maritain, Pierre Teilhard de Chardin, Benedetto Croce and Aldous Huxley. The resulting 260-page volume captured an extremely broad spectrum of theoretical views and justifications for the human rights under consideration.[7]

The replies to the UNESCO enquiry were encouraging. They revealed that the principles underlying the draft Declaration were already present in many cultural and religious traditions, though not always articulated in terms of "rights".[8] For instance, Chinese Confucian philosopher Chung-Shu Lo pointed out in his reply that the absence of formal declarations of rights in China or the difficulties to translate the word "right" into Chinese did not signify "that the Chinese never claimed human rights". He argued that, in fact, "the idea of human rights developed very early in China, and the right of the people to revolt against oppressive rulers was very early established".[9]

Similarly, the Indian political scientist S. V. Puntambekar noted that "great thinkers like Manu and Buddha (. . .) have propounded a code, as it were, of ten essential human

---

6 UNESCO, www.unesco.org/archives/new2010/en/history_of_unesco.html.

7 UNESCO (ed.), *Human Rights: Comments and Interpretations. A Symposium edited by UNESCO with an Introduction by Jacques Maritain*, UNESCO, Paris 1948. Available at: http://unesdoc.unesco.org/images/0015/001550/155042eb.pdf.

8 For a more detailed account, see M. A. Glendon, *A World Made New: Eleanor Roosevelt and the Universal Declaration of Human Rights*, Random House, New York 2002, pp. 73–78.

9 L. Chung-Shu 1948, Human Rights in the Chinese Tradition, in *op. cit.* (pp. 183–187), Human Rights in the Chinese Tradition, in UNESCO (ed.), p. 186.

    L. Chung-Shu 1948, Human Rights in the Chinese Tradition, in UNESCO (ed.), *op. cit.* (pp. 183–187), p. 186.

freedoms and controls or virtues necessary for good life": five social freedoms ("freedom from violence, freedom from want, freedom from exploitation, freedom from violation and dishonour, and freedom from early death and disease") and five individual virtues ("absence of intolerance, compassion or fellow-feeling, knowledge, freedom of thought and conscience and freedom from fear and frustration or despair").[10]

Interestingly, despite the very different philosophical, religious and cultural backgrounds of the respondents to the UNESCO survey, the list of basic rights and values proposed by them were broadly similar. Certainly, the UNESCO philosophers were well aware of the lack of consensus on the ultimate foundations of those rights and values. However, they did not consider these discrepancies as an insurmountable obstacle for an international agreement. In his introduction to the volume that gathered the responses, Jacques Maritain insisted that the goal of the UN efforts was a *practical*, not a *theoretical* one, and pointed out that "agreement between minds can be reached spontaneously, not on the basis of common speculative ideas, but on common practical ideas, not on the affirmation of one and the same conception of the world, of man and of knowledge, but upon the affirmation of a single body of beliefs for guidance in action".[11]

Maritain also reported that in one of the UNESCO meetings someone expressed astonishment that representatives of very different and even opposed ideologies could be able to agree on a list of human rights. The man was told: "Yes, we agree about the rights, but on the condition that no one asks us why".[12]

In line with these remarks, the Committee of experts expressed the view that the issue at stake was not to achieve doctrinal consensus on the foundations of human rights, but to achieve agreement concerning the list of rights that had to be recognized and also concerning the action aiming at the implementation of those rights.[13] From this point of view, and after having examined the responses to the survey, the group concluded in July 1947 that "agreement is possible concerning such a declaration".[14]

The Committee suggested that the agreement should include the following fifteen rights: 1) the right to live, 2) the right of protection of health, 3) the right to work, 4) the right to social assistance in cases of need such as unemployment, infancy, old age, and sickness, 5) the right to property, 6) the right to education, 7) the right to information, 8) the right to freedom of thought and inquiry, 9) the right to self-expression in art and science, 10) the right to justice, which includes the right to fair procedures and freedom from torture and any cruel punishment and illegal arrest, 11) the right to political participation, 12) the right to freedom of speech, assembly, association, press, and religion, 13) the right to citizenship, 14) the right to rebel against an unfair regime, and 15) the right to share in progress.[15]

It is difficult to assess the precise impact that the report of the UNESCO committee had on the Human Rights Commission that drafted the Declaration of Human Rights, but the

---

10 S. V. Puntambekar 1948, Human Freedoms and the Hindu Thinking, in *op. cit.* (pp. 193–197) UNESCO (ed.), p. 195.

11 J. Maritain 1948, Introduction, in *op. cit.* (pp. 4–12) UNESCO (ed.), p. 4.

12 Ibid., p. 4.

13 UNESCO (ed.), *op. cit.*, Appendix II, p. 263.

14 Ibid., p. 15.

15 Ibid., pp. 11–15.

fact is that all the rights proposed by the UNESCO committee were included, in one way or another, in the final version of the instrument adopted in December 1948.[16]

Intersection between human rights and bioethics

Human rights are legal entitlements to have or do something that people have simply by virtue of their humanity. These entitlements concern those basic conditions that are necessary for leading a minimally good life, such as physical and mental integrity, freedom, privacy, health, equal treatment and so on. From this it is not hard to see that there is a close and multifaceted relationship between human rights and medical and health-related issues. It is therefore not surprising that international instruments dealing with bioethics are framed using a rights-based approach. Moreover, they present themselves as an *extension of international human rights law to the field of biomedicine.*[17] In other words, they are not merely ethical or political recommendations, but human rights instruments dealing with a particular kind of issues. This is to say that the new biolegal instruments do not intend to "regulate" or "subsume" bioethics,[18] because bioethics or ethics cannot and should not be regulated by law. They simply stipulate minimal legal standards for promoting respect for human rights in the biomedical context. This is clearly the case of the Universal Declaration on Bioethics and Human Rights (henceforth UDBHR) adopted by UNESCO in 2005.[19]

The importance of the UDBHR lies precisely in the fact that it is the first global legal – though non-binding – instrument that comprehensively addresses the linkage between human rights and bioethics. In this regard, the Chairperson of the drafting group of the Declaration pointed out that the most significant achievement of this document consists precisely in having integrated the bioethical analysis into a human rights framework.[20] As noted by the Explanatory Memorandum to the Preliminary Draft Declaration, "the drafting group also stressed the importance of taking international human rights legislation as the essential framework and starting point for the development of bioethical principles."[21] The Explanatory Memorandum also points out that there are two broad streams at the origin of the norms dealing with bioethics. The first one can be traced to antiquity, in particular to Hippocrates, and is derived from reflections on the practice of medicine. The second one, conceptualized in more recent times, has drawn upon the developing international human rights law. Furthermore, the Memorandum states: "One of the important

---

16 The right to rebellion is the only one that does not appear in the body of the UDHR, but it can be inferred from its Preamble, which states: "Whereas it is essential, if man is not to be compelled to have recourse, as a last resort, to rebellion against tyranny and oppression, that human rights should be protected by rule of law".

17 R. Andorno, *Principles of International Biolaw: Seeking Common Ground at the Intersection of Bioethics and Human Rights*, Bruylant, Brussels 2013, p. 17.

18 T. Faunce, Will International Human Rights Subsume Medical Ethics? Intersections in the UNESCO Universal Bioethics Declaration, in *Journal of Medical Ethics*, 2005, vol. 31, pp. 173–178.

19 Another example of this trend is the Council of Europe's Convention on Biomedicine and Human Rights (1997).

20 M. Kirby, UNESCO and Universal Principles in Bioethics: What's Next?, in UNESCO (ed.), *Twelfth Session of the International Bioethics Committee (IBC). December 2005. Proceedings*, UNESCO, Paris 2006, p. 126.

21 UNESCO, *Explanatory Memorandum on the Elaboration of the Preliminary Draft Declaration on Universal Norms on Bioethics*, UNESCO, Paris 2005, para 11.

achievements of the declaration is that it seeks to unite these two streams. It clearly aims to establish the conformity of bioethics with international human rights law."[22]

Certainly, the UDBHR, like any intergovernmental declaration, belongs to the category of soft law instruments, which are weaker than conventions because they are not immediately binding for states (this topic will be widely discussed in Part II of the volume). The use of soft law instruments has rapidly developed over the last decades as a tool for dealing with sensitive matters such as human rights, the protection of the environment and bioethical issues. However, it would be misleading to deduce from the non-binding effect of soft law instruments that they are purely rhetorical statements and are therefore deprived of any legal effects. In reality, although they are not immediately binding for states, they are *potentially binding* in the sense that they are thought of as the beginning of a gradual process in which further steps are needed to make of such agreements binding rules for states. It is also worth mentioning that, in the course of time, soft law standards may become binding rules in the form of customary law and jurisprudential criteria, as it happened with the Universal Declaration of Human Rights of 1948.[23] It is not to exclude that a similar process of hardening into binding rules could take place in the coming decades with the UDBHR.

What are the reasons for resorting to human rights for developing normative frameworks relating to bioethics? There is, first of all, a historical reason. Both international human rights law and modern medical ethics were born together as a response to the dramatic events of the Second World War. Several provisions of the Universal Declaration of Human Rights were informed by the horror caused by the revelation that prisoners of concentration camps were used as subjects of brutal medical experiments. This same shocking discovery led the Nuremberg trial to develop the famous ten principles for medical research, better known as the "Nuremberg Code". In this regard, it has been said that the Second World War was "the crucible in which both human rights and bioethics were forged, and they have been related by blood ever since."[24]

In addition to this historical common ground, the close link between bioethics and human rights can also be explained by the circumstance that medical activities are directly related to some of the most important human rights (the right to life, the right to physical integrity, to confidentiality of personal data, the right to health care, etc.). Therefore, it is understandable that both fields overlap to some extent with each other, and that the already existing human rights framework is used to ensure the protection of such basic human goods, also in the field of biomedicine.

There is also a very practical reason for integrating bioethical standards into a human rights framework: there are few, if any, mechanisms available other than human rights to function as a global normative foundation in biomedicine. The human rights framework provides "a more useful approach for analysing and responding to modern public health challenges than any framework thus far available within the biomedical tradition."[25] Similarly, it has been argued that the recourse to human rights is fully justified on the ground that bioethics suffers from the plurality and range of actors involved and the overproduc-

---

22 Ibid., para 12.
23 R. Andorno, Human Dignity and Human Rights as a Common Ground for a Global Bioethics, in *Journal of Medicine and Philosophy*, 2009, vol. 34, no. 3, pp. 223–240.
24 G. J. Annas, *American Bioethics: Crossing Human Rights and Health Law Boundaries*, Oxford University Press, New York 2005, p. 160.
25 J. Mann, Health and Human Rights: Protecting Human Rights Is Essential for Promoting Health, in *British Medical Journal*, 1996, vol. 312, pp. 924–925.

tion of divergent norms, while "human rights offers a strong framework and a common language, which may constitute a starting point for the development of universal bioethical principles".[26] According to Richard Ashcroft, casting bioethical issues into human rights terms allows "a well-tested and long-established common language, rhetoric and institutional practice" to better identify the problems at stake and ideally, to find the possible solutions to them.[27] This is to say that, while bioethics suffers from the plurality of actors and divergent theories, human rights offers a strong, effective and enforceable set of standards.

Another reason explaining the recourse to human rights in standard-setting action in bioethics is the *universal* nature of human rights, because such universality facilitates the development of global standards for biomedicine. Human rights are, by definition, conceived as transcending cultural diversity and national boundaries. They are held to be universal in the sense that "all people have and should enjoy them, and to be independent in the sense that they exist and are available as standards of justification and criticism whether or not they are recognized and implemented by the legal system or officials of a country".[28] In such a sensitive field as bioethics, where diverse socio-cultural, philosophical and religious traditions come into play, this universality is a very precious asset when formulating global standards for biomedical issues. This universality should however not be understood as a rigid one. Human rights are regarded by international law as flexible enough to be compatible, within certain limits, with respect for cultural diversity. As noted by Jack Donnelly, the human rights system allows local variations, not in the substance, but in the *form* in which particular rights are interpreted and implemented.[29]

UNESCO's involvement in bioethics

UNESCO has acted as a pioneer in standard-setting action in global bioethics. Surprisingly, its involvement in this field has been harshly criticized by some scholars. For instance, during the drafting work of the Universal Declaration on Bioethics and Human Rights, it was argued that UNESCO would be in an "obvious attempt at meddling in the professional domain of another United Nations (UN) agency, WHO" and that "it is entirely unclear why UNESCO should concern itself with such a matter".[30] In the same vein, it was claimed that "UNESCO is clearly overstepping its mandate and encroaching on that of the World Health Organization (WHO)".[31]

More recently, Aurora Plomer has levelled similar objections to UNESCO's initiatives on bioethics. In her view, "UNESCO's more recent standard-setting activities in bioethics (. . .) look somewhat detached from UNESCO's own programs to support the advancement and

---

26 H. Boussard, The 'Normative Spectrum' of an Ethically-Inspired Legal Instrument: The 2005 Universal Declaration on Bioethics and Human Rights, in Francesco Francioni (ed.), *Biotechnologies and International Human Rights*, Hart Publishing, Oxford 2007, p. 114.

27 R. Ashcroft, Could Human Rights Supersede Bioethics?, in *Human Rights Law Review*, 2010, vol. 10, no. 4, p. 644.

28 J. Nickel, *Making Sense of Human Rights: Philosophical Reflections on the Universal Declaration of Human Rights*, University of California Press, Berkeley 1987, p. 561.

29 J. Donnelly, *Universal Human Rights in Theory and Practice*, Cornell University Press, Ithaca, NY 1989, pp. 109–142.

30 U. Schuklenk, W. Landman, From the Editors, in *Developing World Bioethics*, 2005, vol. 5, no. 3, pp. 3–4.

31 J. Williams, UNESCO's Proposed Declaration on Bioethics and Human Rights: A Bland Compromise, in *Developing World Bioethics*, 2005, vol. 5, no. 3, pp. 210–215.

diffusion of science, overly focused on health/biomedical concerns which fall more squarely within WHO's mandate and overly preoccupied with limitations on modern biotechnologies on which there is no clear global consensus".[32]

A variety of answers to these criticisms can be raised.[33] First of all, the division of competences between UN agencies is not as clear-cut as it might seem at first glance. This is especially the case in issues that are at the intersection of different disciplines, like those relating to bioethics. Therefore, when interdisciplinary matters are at stake, extreme caution is needed before accusing a UN agency of overstepping its mandate. There is no doubt that the World Health Organization (WHO), which is the specialized UN agency for health, is called to play a major role in the standard-setting activities in the field of public health. Nevertheless, as a scholar has pointed out, the WHO cannot manage this task alone, and this for several reasons: first, the field is growing rapidly encompassing more diverse and complex concerns, due to its interdisciplinary nature; second, the WHO has very limited experience in international health lawmaking; third, such a task would deplete the organization's limited resources and undermine its ability to fulfill its well-established and essential international health functions; fourth, member states are highly unlikely to limit their autonomy and freedom by granting to the WHO alone such an expansive new mandate; fifth, decentralization of the international lawmaking enterprise presents great advantages that cannot be ignored.[34]

Furthermore, it is unclear why the only UN agency specialized in *sciences* (both natural and human sciences) and having served for decades as a forum for *philosophical* discussion on cross-cultural issues should abstain from making any contribution to the global normative guidance of biomedical sciences. It must be reminded that the purpose of UNESCO is, according to its Constitution, to promote "collaboration among nations through education, science and culture in order to further universal respect for justice, for the rule of law and for the human rights and fundamental freedoms".[35] Hence, it is understandable that an organization with such an ambitious mission may consider its duty to make its own contribution to the development of human rights standards in the field of biomedicine.

It should be also reminded that, after all, UNESCO and WHO are composed of the same member states. Therefore, any conflict of competences between these two bodies is to some extent meaningless. But there is a more substantial reason for favouring simultaneous involvement of both UN agencies in the field of bioethics: their standard-setting activities operate at *different levels*. While UNESCO tends to produce general normative frameworks of a predominantly philosophical and legal nature, WHO's guidelines are usually more technical and focused on very specific health-related issues. Thus, since the approach followed by both organizations is different, their respective involvement in this matter is not necessarily overlapping, but can perfectly coexist.

In addition, it is noteworthy that UNESCO has a long experience in standard-setting. At least since the 1970s, it has been involved in the development of around 28 international conventions, 13 declarations and about 33 recommendations relating to science, education and human rights, including the Convention against Discrimination in Education (1960),

32 A. Plomer, *op. cit.*, p. 161.
33 See R. Andorno, Global Bioethics at UNESCO: In Defence of the Universal Declaration on Bioethics and Human Rights, in *Journal of Medical Ethics*, 2007, vol. 33, no. 3, pp. 150–154.
34 A. L. Taylor, Governing the Globalization of Public Health, in *Journal of Law, Medicine and Ethics*, 2004, vol. 32, no. 3, pp. 500–508.
35 UNESCO Constitution, 16 November 1945 (Article 1).

the Universal Copyright Convention (1971), the Declaration on Principles of International Cultural Cooperation (1966), the Declaration on Race and Racial Prejudice (1978), the Declaration on the Responsibilities of the Present Generations towards Future Generations (1997), the Recommendation on the Status of Scientific Researchers (1974), the Recommendation Concerning the International Standardization of Statistics on Science and Technology (1978) and the Convention on the Protection and Promotion of the Diversity of Cultural Expressions (2005). Thus, even for merely pragmatic reasons, it is difficult to see why the international community could not take advantage of this long experience regarding sciences, its cross-cultural impact and its significance for human rights in order to set up global bioethical standards.

Furthermore, UNESCO's involvement in bioethics did not just start in 2005, but dates back to the 1970s, when this organization began to organize symposia and conferences on bioethical issues, mainly related to genetics, life sciences and reproductive technologies.[36] In 1993, Federico Mayor, then Director-General of UNESCO, decided to set up an International Bioethics Committee (IBC) composed by experts from different countries and disciplines. The first task of the IBC was to prepare the preliminary draft of the Universal Declaration on the Human Genome and Human Rights, which was adopted in 1997. Thereafter, the IBC worked on the drafting of the International Declaration on Human Genetic Data, finalized in 2003, and of the Universal Declaration on Bioethics and Human Rights, adopted by UNESCO's General Conference in 2005. Since its creation, the IBC produced around 22 reports on a variety of bioethical topics such as genetic screening (1994), genetic counseling (1995), ethics and neurosciences (1995), the ethics of experimental treatments (1996), confidentiality and genetic data (2000), embryonic stem cells (2001), the ethics of intellectual property and genomics (2002), preimplantation genetic diagnosis and germ-line interventions (2003), informed consent (2008), human cloning (2009), social responsibility and health (2010), human vulnerability (2013) and benefit-sharing (2015). The truth is that no other intergovernmental organization could claim the same level of experience at the intersection of life sciences, ethics and human rights.[37]

In addition to the decades-long involvement of UNESCO in bioethical issues, another truth is that a conflict of competences between two or more UN agencies about the governance of life sciences would be quite pointless. Such a conflict would be as absurd as a dispute between philosophers, physicians and lawyers over the "ownership" of bioethics. Obviously, none of these disciplines has the monopoly of bioethics. Since this field is by its very nature an interdisciplinary domain, all related professions (and likewise, all related UN bodies) have the right – and the duty – to make their specific contribution to this complex area.

The Universal Declaration of Bioethics and Human Rights

On 19 October 2005, the Universal Declaration on Bioethics and Human Rights was adopted at the UNESCO's General Conference by representatives of 191 countries. The overall goal of this instrument is "to provide a universal framework of principles and procedures to guide States in the formulation of their legislation, policies or other instruments in the field of bioethics" (Article 2a).

---

36 H. ten Have, M. Jean, Introduction, in Henk ten Have and Michèle Jean (eds.), *The UNESCO Declaration on Bioethics and Human Rights: Background, Principles and Application*, UNESCO, Paris 2009, p. 23.
37 A. L. Taylor, *op. cit.*

The drafting of the UDBHR is the result of a combined effort by bioethics experts from different countries and disciplines who sat on the International Bioethics Committee (IBC) of UNESCO, and governmental representatives of UNESCO member states. The drafting work was preceded by a report of an IBC working group that assessed the feasibility of such an instrument. The group, chaired by Leonardo De Castro (Philippines) and Giovanni Berlinguer (Italy), concluded by supporting the initiative and affirming the need to develop "a worldwide common sense in order to foster understanding and cohesion in relation to new ethical categories and new practical possibilities emerging from science and technology".[38] Encouraged by these conclusions, the IBC, chaired at the time by Michèle Jean (Canada), prepared the preliminary draft declaration, after almost two years of discussions and public consultations with governmental and non-governmental organizations. Justice Michael Kirby (Australia) chaired the drafting group, which was open to all IBC members. To ensure transparency in the process, the successive versions of the document were posted on the Internet as they were being developed. In January 2005, the draft was examined by the Intergovernmental Bioethics Committee (IGBC) and, finally, it was revised in two successive meetings of governmental representatives, who introduced several amendments.[39]

Despite the great number of existing international guidelines, statements and declarations relating to bioethics, the UDBHR made its own specific contribution to this field. First, because it is a *legal*, and not a merely ethical, instrument like those produced by non-governmental organizations (for instance, the World Medical Association). Second, because it is not restricted to a particular area of bioethics, but provides a comprehensive framework of principles for all biomedical activities. Third, because it is the first global intergovernmental instrument that addresses the linkage between human rights and bioethics.

Section II of the Declaration sets out 16 substantive principles relating to bioethics. These principles are to be understood as "complementary and interrelated" (Article 26). This means that the relationship between them is non-hierarchical.[40] The complexity of bioethical dilemmas in real life makes it impossible to establish in abstract terms a clear priority of some principles over others. Therefore, in case of conflict between two or more principles, the priority of one of them will be determined taking into account the particular circumstances of each case, as well as the cultural specificities of each society. However, this does not preclude that the principle of respect for human dignity, due to its inescapable overarching nature, will always play a role in every bioethical decision.

The principles proclaimed in the Declaration are the following:

– Respect for human dignity and human rights (Article 3.1)
– Priority of the individual's interests and welfare over the sole interest of science or society (Article 3.2)
– Beneficence and non-maleficence (Article 4)
– Autonomy (Article 5)
– Informed consent (Article 6)
– Protection of persons unable to consent (Article 7)
– Special attention to vulnerable persons (Article 8)

38 UNESCO IBC, *Report on the Possibility of Elaborating a Universal Instrument on Bioethics*, 13 June 2003. Available at: http://unesdoc.unesco.org/images/0013/001302/130223e.pdf.
39 See a detailed account of the drafting process in H. ten Have and M. Jean, ibid., pp. 17–55.
40 E. Gefenas, Article 26: Interrelation and Complementarity of the Principles, in H. ten Have and M. Jean (eds.), ibid., pp. 327–333.

- Privacy and confidentiality (Article 9)
- Equality, justice and equity (Article 10)
- Non-discrimination and non-stigmatization (Article 11)
- Respect for cultural diversity and pluralism (Article 12)
- Solidarity and cooperation (Article 13)
- Access to health care and essential medicines (Article 14)
- Benefit sharing (Article 15)
- Protection of future generations (Article 16)
- Protection of the environment, the biosphere and biodiversity (Article 17)

Section III of the Declaration, entitled "Application of the Principles", is devoted to principles of a more procedural nature such as:

- The requirement for professionalism, honesty, integrity and transparency in the decision-making process regarding bioethical issues (Article 18)
- The need to establish independent, multidisciplinary and pluralist ethics committees (Article 19)
- The call for an appropriate risk assessment and management in the biomedical field (Article 20)
- The need for justice in transnational research (Article 21)

It could be objected that most of the above-mentioned principles are not completely new, as they can be found in previous international human rights instruments. This is true, but this does not render the Declaration useless or redundant. On the contrary, it is precisely the accumulation, convergence and complementarity of principles that gradually shape social consciousness in the international community, both internationally and transnationally.[41] Besides that, the greatest merit of the Declaration does not lie in having "invented" entirely new human rights principles, but in having developed them from previous international instruments to adapt them to the specific field of biomedicine, and in having assembled them together into a single, coherent legal instrument.

> It could be also argued that the Declaration's provisions are too vague and ambiguous. Indeed, the principles it proclaims are couched in very general terms and the document offers little guidance about their precise meaning and implications. This vagueness has led some scholars to argue that the Declaration will not provide much guidance because everyone can interpret the text as they like.[42] In reality, the importance of laying down general principles, even if they might appear to be too vague, should not be underestimated. They are meant to serve as a starting point for further discussion and negotiation that could lead to more precise regulations, especially at a national level. It should not be forgotten that *national governments*, not international organizations, are the primary agents for the realization of human rights. The international community has an important role to play in setting up widely accepted standards, but once these

---

41 P. M. Dupuy, The Impact of Legal Instruments Adopted by UNESCO on General International Law, in Abdulqawi Yusuf (ed.), *Normative Action in Education, Science and Culture: Essays in Commemoration of the Sixtieth Anniversary of UNESCO*, UNESCO, Paris 2007, p. 356.

42 D. Benatar, The Trouble with Universal Declarations, in *Developing World Bioethics*, 2005, vol. 5, no. 3, pp. 220–224.

principles have been set up, the primary locus for their implementation is within each state. In fact, legal principles, such as human rights, are necessarily very general and leave always open the possibility of various interpretations. In addition, the criticism of vagueness rests on a misunderstanding of the nature and scope of this kind of instrument. Intergovernmental declarations like the UDBHR should not be assessed with purely academic criteria, because they are not the product of purely scholarly work, but rather a kind of compromise between a theoretical conceptualization made by experts and what is practically achievable given the political choices of governments.[43]

In this regard, it is also helpful to remind here the distinction between "rules" and "principles" made by legal philosophers.[44] "Rules" are norms that are applicable in an all-or-nothing fashion (for instance, "it is forbidden to drive faster than 50 km/h in the urban area"). If the facts stipulated in a rule are given, then either the rule is valid, in which case the answer it supplies must be accepted, or it is not, in which case it contributes nothing to the decision. But this is not the way principles operate. Principles alone never completely determine the content of a particular decision. They are "optimization commands" (*Optimierungsgebote*), which can be carried out to different degrees depending on the circumstances.[45] Human rights are one of the best examples of "principles". Principles are always valid, but they have a dimension of *weight* or *importance*. When principles intersect, one who must resolve the conflict has to take into account the relative weight of each one. This is perfectly applicable to the norms contained in the UDBHR, which are, technically speaking, "principles", not "rules".

The very general nature of many of the norms included in the Declaration can also be explained for practical reasons. Indeed, it would have been impossible to reach a global agreement on the precise meaning and justification of fundamental moral notions such as "human dignity", "autonomy", "justice", "benefit", "harm" or "solidarity", which have a long philosophical history and are, to some extent, conditioned by cultural factors. The IBC members, who were involved in the drafting of the UDBHR, were well aware of the impossibility of finding a common theoretical justification for the principles to be included in the document. Here, a parallel can be drawn with the conclusions of the committee of intellectuals consulted by UNESCO in 1947, which was mentioned previously. In spite of the very different and even opposing philosophical and socio-cultural backgrounds of the IBC members, they succeeded to agree on a list of principles to be incorporated into the UDBHR. Paraphrasing the quotation by Maritain cited previously, the IBC members could have perfectly said: "We all agree about the principles applicable to bioethics, but on the condition that no one asks us why".

Conclusions

UNESCO inherited the hopeful legacy of humanistic philosophy that emerged in the aftermath of the Second World War. Since its creation in 1945, the Organization has struggled

---

43 F. Baylis, Global Norms in Bioethics: Problems and Prospects, in R. Green, A. Donovan, S. Jauss (eds.), *Global Bioethics. Issues of Conscience for the Twenty-First Century*, Clarendon Press, Oxford 2008, pp. 323–339.

44 See R. Dworkin, *Taking Rights Seriously*, Duckworth, London 1977, pp. 22–28.

45 R. Alexy, *Theorie der Grundrechte*, Suhrkamp, Frankfurt 1994, pp. 71–77.

to consolidate its overwhelming sense of purpose as well as to preserve the strength of the concept of humanity and its expression in international law.[46]

Being the "intellectual agency" of the United Nations, UNESCO perceived well from the very beginning the need to anticipate the theoretical problems that may arise when drafting an international declaration of human rights. In the decades that followed, the Organization continued contributing to the development of human rights through its standard-setting action, in particular regarding those rights related to education, culture and science. It is very significant that, when addressing bioethical issues, which are closely related to the cultural specificity of each society, UNESCO did not abandon the conviction, well enshrined in international law, that human rights are universal and therefore transcend cultural and political boundaries. This means that people are entitled to basic rights simply by virtue of their humanity, and irrespective of ethnic origin, sex, nationality, religion, or social and economic status. The circumstance that bioethical issues are inevitably linked to the deepest socio-cultural and religious values of every society was not regarded by UNESCO as an obstacle to the formulation of universal principles in bioethics. With the adoption of the UDBHR in 2005, UNESCO has continued to consolidate its mission of promoting collaboration among nations through education, science and culture. Once again, universal human rights and their grounding in human dignity are called to contribute to peace by playing a role as a bridge between cultures.

Bibliography

Andorno, Roberto 2007, Global Bioethics at UNESCO: In Defence of the Universal Declaration on Bioethics and Human Rights, *Journal of Medical Ethics*, 33: 150–154.

Andorno, Roberto 2009, Human Dignity and Human Rights as a Common Ground for a Global Bioethics, *Journal of Medicine and Philosophy*, 34: 223–240.

Andorno, Roberto 2013, *Principles of International Biolaw: Seeking Common Ground at the Intersection of Bioethics and Human Rights*, Brussels: Bruylant.

Annas, George J. 2005, *American Bioethics: Crossing Human Rights and Health Law Boundaries*, New York: Oxford University Press.

Ashcroft, Richard 2010, Could Human Rights Supersede Bioethics?, *Human Rights Law Review*, 10(4): 639–660.

Baylis, Françoise 2008, Global Norms in Bioethics: Problems and Prospects, in *Global Bioethics: Issues of Conscience for the Twenty-First Century* (pp. 326–339) R. Green, A. Donovan, and S. Jauss, eds., Oxford: Clarendon Press.

Benatar, David 2005, The Trouble with Universal Declarations, *Developing World Bioethics*, 5: 220–224.

Berlinguer, Giovanni, De Castro, Leonardo 2003, *Report of the IBC on the Possibility of Elaborating a Universal Instrument on Bioethics*, UNESCO 13.06.2003, SHS-2004/DECLAR.BIO-ETHIQUE CIB/6.

Boussard, Hélène 2007, The 'Normative Spectrum' of an Ethically-Inspired Legal Instrument: The 2005 Universal Declaration on Bioethics and Human Rights, in *Biotechnologies and International Human Rights* (pp. 97–127) Francioni, Francesco (ed.), Oxford: Hart Publishing, 97–127.

Bowen, Matthew D. 2008, Race, Religion and Informed Consent: Lessons from Social Science, *Journal of Law, Medicine & Ethics*, 36: 150.

46 P. M. Dupuy, *op. cit.*, p. 362.

Bowman, Kerry 2004, What Are the Limits of Bioethics in a Culturally Pluralistic Society?, *Journal of Law, Medicine & Ethics*, 32: 664–669.

Candib, Lucy 2002, Truth Telling and Advance Planning at the End of Life: Problems with Autonomy in a Multicultural World, *Family Systems & Health*, 20: 213–228.

Cheng-the Tay, Mark, Sung, Lin 2001, Developing a Culturally Bioethics for Asian People, *Journal of Medical Ethics*, 27(1): 51–54.

D'Agostino, Francesco 1998, *Bioetica, nella prospettiva della filosofia del diritto*, Torino: Giappichelli.

Davidson, Scott 2001, Human Rights, Universality and Cultural Relativity: In Search of a Middle Way, *Human Rights Law and Practice*, 6: 97.

de Varennes, Fernand 2006, The Fallacies in the "Universalism versus Cultural Relativism" Debate in Human Rights Law, *Asia Pacific Journal on Human Rights and the Law*, 7(1): 67–84.

Dundes Renteln, Alison 1990, *International Human Rights: Universalism versus Relativism*, New York: Sage Publications.

Durkheim, Emile 1965, *Elementary Forms of Religious Life*, New York: Free Press.

Fan, Rong 1997, A Report from East Asia: Self-Determination vs. Family-Determination: Two Incommensurable Principles of Autonomy, *Bioethics*, 11: 309–322.

Faunce, Thomas 2005, Will International Human Rights Subsume Medical Ethics? Intersections in the UNESCO Universal Bioethics Declaration, *Journal of Medical Ethics*, 31: 173–178.

Faunce, Thomas 2014, *Bioethics and Human Rights*, in *Handbook of Global Bioethics* (pp. 467–484) ten Have, Henk, Gordijn, Bert, eds., Netherlands: Springer, 467–484.

Finnis, John 1980, *Natural Law and Natural Rights*, Oxford: Clarendon Press.

Gaudreault-DesBiens, Jean-François 2009, Religious Challenges to the Secularized Identity of an Insecure Polity: A Tentative Sociology of Québec's 'Reasonable Accommodation' Debate, in *Legal Practice and Cultural Diversity* (pp. 151–274) Grillo, Ralph, ed., Farnham: Ashgate.

Glendon, Mary Anne 2002, *A World Made New: Eleanor Roosevelt and the Universal Declaration of Human Rights*, New York: Random House.

Gordon, Edwin 1995, Multiculturalism in Medical Decision-Making: The Notion of Informed Waiver, *Fordham Urban Law Journal*, 23(4).

Hatch, Elvin 1983, *Culture and Morality*, New York: Columbia University Press.

Helman, Cecil H. 2007, *Culture, Health and Illness*, London: Hodder Arnold Press.

Kirby, Michael 2006, UNESCO and Universal Principles in Bioethics: What's Next?, in *Twelfth Session of the International Bioethics Committee (IBC): December 2005. Proceedings* UNESCO, ed., Paris: UNESCO, 121–136.

Kirby, Michael 2009a, *Article 1: Scope, in The UNESCO Universal Declaration of Human Rights: Backgrounds, Principles and Application* (pp. 67–81), Paris: UNESCO Publishing.

Kirby, Michael 2009b, Human Rights and Bioethics: The Universal Declaration of Human Rights and UNESCO Universal Declaration of Bioethics and Human Rights, *Journal of Contemporary Health Law & Policy*, 25: 309.

Langlois, Adele 2013, *Negotiating Bioethics: The Governance of UNESCO's Bioethics Programme*, London: Routledge.

Legg, Andrew 2012, *The Margin of Appreciation in International Human Rights Law: Deference and Proportionality*, Oxford: Oxford University Press.

Lévinas, Emmanuel 1990, Ethique et transcendance. Entretiens avec le philosophe Emmanuel Lévinas, in *Médicine et éthique: Le dévoir d'humanité* (p. 475) Hirsch, Emmanuel (ed.), Paris: Le Cerf.

Logan, William S. 2007, Closing Pandora's Box: Human Rights Conundrums in Cultural Heritage Protection, in *Cultural Heritage and Human Rights* (pp. 33–52) Silverman, Helaine, Ruggles, D. Fairchild, eds., New York: Springer.

Loustaunau, Martha O., Sobo, Elisa J. 1997, *The Cultural Context of Health, Illness, and Medicine*, Westport, CT: Bergin & Garvey.

MacDonald, Ronald St. J. 1993, *The Margin of Appreciation, in The European System for the Protection of Human Rights* (ed.) MacDonald, Ronald St. J., Matscher, Franz, Petzold, Herbert, Dordrecht: M. Nijhoff.

Macioce, Fabio 2014, *Il Nuovo Noi. Immigrazione e integrazione come problemi di giustizia*, Torino: Giappichelli.

Macioce, Fabio 2016a, A Multifaceted Approach to Legal Pluralism and Ethno-Cultural Diversity, *The Journal Of Legal Pluralism And Unofficial Law*, 48(1): 1–16.

Macioce Fabio 2016b, Balancing Cultural Pluralism and Universal Bioethical Standards: A Multiple Strategy, *Medicine, Health Care and Philosophy*, 19: 393–402.

Macklin, Ruth 2003, Dignity Is a Useless Concept, *British Medical Journal*, 327, 1419–1420.

Macklin, Ruth 2005, Yet another Guideline? The UNESCO Draft Declaration, *Developing World Bioethics*, 5: 244–250.

Macklin, Ruth 2009, Global Health, in *Oxford Handbook of Bioethics*, Oxford: Oxford University Press.

Macmillan, Margaret 2003, *Paris 1919: Six Months That Changed the World*, New York: Random House.

Marshall, Patricia 2000, Informed Consent in International Health Research: Cultural Influences on Communication, in *Biomedical Research Ethics: Updating International Guidelines: A Consultation* (pp. 100–134) Levine, R. J., Gorovitz, S., Gallagher, J., eds, Geneva, Switzerland 15–17 March, 2000. Geneva: Council for International Organizations of Medical Sciences.

Marshall, Patricia, Koenig, Barbara 2004, Accounting for Culture in a Globalized Bioethics, *Journal of Law, Medicine and Ethics*, 32: 252–266.

Nickel, James 1987, *Making Sense of Human Rights: Philosophical Reflections on the Universal Declaration of Human Rights*, Berkeley, CA: University of California Press.

Parekh, Bhikhu 2006, *Rethinking Multiculturalism: Cultural Diversity and Political Theory*, London: Macmillan.

Plomer, Aurora 2015, *Patents, Human Rights and Access to Science*, Cheltenham: Edward Elgar Publishing.

Pogge, Thomas W. 2001, Priorities of Global Justice, in *Global Justice* (pp. 312–331) Pogge, Thomas (ed.), Oxford: Blackwell Publishers.

Rawls, John 1993, *Political Liberalism*, New York: Columbia University Press.

Revel, Michel 2009, Article 12: Respect for Cultural Diversity and Pluralism, in *The UNESCO Universal Declaration of Human Rights, Backgrounds, Principles and Application* (pp. 199–209), Paris: UNESCO Publishing.

Sabatello, Maya 2009, *Children's Bioethics: The International Biopolitical Discourse on Harmful Traditional Practices and the Right of the Child to Cultural Identity*, Leiden, NLD: Martinus Nijhoff.

Shachar, Ayelet 2001, *Multicultural Jurisdictions: Cultural Differences and Women's Rights*, Port Chester, NY: Cambridge University Press.

Shaibu, Sheila 2007, Ethical and Cultural Considerations in Informed Consent in Botswana, *Nursing Ethics*, 14(4): 503–509.

Smith, George P. 2005, Human Rights and Bioethics: Formulating a Universal Right to Health, Health Care, or Health Protection?, *Vanderbilt Journal of Transnational Law*, 38: 1295.

Snow, Loudell F. 1974, Folk Medical Beliefs and Their Implications for Care Patients: A Review Based on Studies among Black Americans, *Annals of International Medicine*, 81(1): 82–96.

Sunstein, Cass R. 1995, Incompletely Theorized Agreements, *Harvard Law Review*, 108(7) (May, 1995): 1733–1772.

Sweet, William, Masciulli, Joseph 2011, Biotechnologies and Human Dignity, Bulletin of Science, *Technology & Society*, 31(1): 6–16.

Taylor, Allyn L. 2004, Governing the Globalization of Public Health, *Journal of Law, Medicine and Ethics*, 32(3): 500–508.

ten Have, Henk 2013, Global Bioethics: Transnational Experiences and Islamic Bioethics, *Zygon*, 48(3): 600–617.

ten Have, Henk, Gordijn, Bert 2013, Global Bioethics, in *Compendium and Atlas of Global Bioethics* (pp. 3–18) ten Have, Henk, Gordijn, Bert (eds.), Berlin: Springer.

ten Have, Henk, Jean, Michèle S. (eds.) 2009, Introduction, in *The UNESCO Universal Declaration of Human Rights: Backgrounds, Principles and Application* (pp. 18–56), Paris: UNESCO Publishing.

Teson, Fernando 1985, International Human Rights and Cultural Relativism, *Virginia Journal of International Law*, 25: 869–886.

Thomasma, David C. 1997, Bioethics and International Human Rights, *Journal of Law, Medicine & Ethics*, 25: 295–306.

Thomasma, David C. 2008, Evolving Bioethics and International Human Rights, in *Autonomy and Human Rights in Health Care: An International Perspective* (pp. 11–24) Weisstub, David N., Díaz Pintos, Guillermo, eds., Netherlands: Springer.

Tilley, Jack 2000, Cultural Relativism, *Human Rights Quarterly*, 22: 501–47.

Turner, Leigh 2001, Medical Ethics in a Multicultural Society, *Journal of the Royal Society of Medicine*, 94: 592.

Valderrama, Fernando 1995, *A History of UNESCO*, Paris: UNESCO Publishing.

Walzer, Michael 1994, *Thick and Thin: Moral Argument at Home and Abroad*, London: University of Notre Dame Press.

II Adoption and implementation of the Universal Declaration on Bioethics and Human Rights (UDBHR)

An important step in the construction of Global Bioethics

Michèle Stanton-Jean

Introduction

For centuries, nation States were sovereign on their own soil, with absolutely no inclination to surrender their sovereignty in any regard or in any way. They developed and managed their own governments, laws, policies, wars, regulations and people as they chose, in line with their own values, history and culture. But over time, with developments in knowledge, science, technology and the global economy, nation states have been forced, not without some difficulty, to move away from this territory-centered approach.

The creation, after the Second World War, of the network of United Nations organizations like the World Health Organization (WHO), the Food and Agriculture Organization (FAO) and the United Nations Educational, Scientific and Cultural Organization (UNESCO), among others, was in a sense a signal sent to countries that the time had come to talk to each other and to take common positions on issues covered by the different mandates of these organizations. Emerging from two terrible wars, the newly created organizations agreed that international discussions might well be a better way to hear how 'others' were thinking about issues than going to war.

The UNESCO Constitution came into force on November 4, 1946. Its preamble provided a clear rationale for the mandate of the Organization: "That since wars begin in the minds of men, it is in the minds of men that the defences of peace must be constructed; That ignorance of each other's ways and lives has been a common cause, throughout the history of mankind, of that suspicion and mistrust between the peoples of the world through which their differences have all too often broken into war" (UNESCO 1945).

The UNESCO Constitution was instrumental in articulating and developing this common good approach. The architects of the Organization gave it a mandate to "Build peace in the minds of men" by defining how nations could come together to discuss education, culture and science. Article I, Paragraph 2 states that the Organization will "recommend such international agreements as may be necessary to promote the free flow of ideas by word and image" (UNESCO 1945).

Article IV, Paragraph B4 identifies two categories of instruments that can be developed by the Organization: "Conventions and Recommendations", both of which have to be approved by the General Conference. Another category, "Declarations", was also identified in the Constitution and has become quite widely used in recent years (UNESCO 2007a). Declarations are similar to Recommendations but the choice of this appellation reflects their importance. They are solemnly adopted by General Conferences and, while they remain non-binding (soft law) instruments, Declarations engage governments to implement them in their own particular countries. They also constitute tools that civil society and the scientific community can use to persuade decision-making authorities to take these positions into consideration when they draft policies and legislation (Stanton-Jean 2016). Many Conventions, Declarations and Recommendations have been developed over the years to advance this agenda, thus fulfilling the standard-setting mission of UNESCO.

UNESCO and ethics

The standard-setting mission of UNESCO included ethics from the outset. As former Director-General Koïchiro Matsuura wrote:

> From the beginning of the Organization's activities in this field [Bioethics], UNESCO's General Conference decided to adopt a gradual and prudent approach based on the knowledge available on this complex subject matter, which lies at the interface of many disciplines. Furthermore it decided to take into account the diverse contexts (scientific, cultural, social and economic), in which ethical thinking unfolds in different parts of the world. This approach has led to two important consequences. The first is the use of the "Declaration" rather than a convention or recommendation for the setting of standards in the field of bioethics. The second consequence is the articulation of broad principles and norms, which could be accepted by all Member States of UNESCO in view of the universal nature of the issues involved.
>
> (UNESCO 2007b)

The quotation from Matsuura provides a valuable summary of the Bioethics philosophy of UNESCO, based on an approach designed to cautiously achieve cumulative progress, grounded in good science, and requiring the involvement of committees of experts in the field from different regions and professional backgrounds: The International Bioethics Committee (IBC) and the Intergovernmental Bioethics Committee, made up of representatives of governments (IGBC).

The fact that one committee (the IBC) was free to put forward proposals based on the best scientific knowledge available at the time the Declaration was being drafted, while members of the other committee, although concerned by the scientific issues, had to take positions reflecting the positions of their own governments and the social and economic policies, goals and situation of those governments, made the interaction between the two committees somewhat challenging. At the same time, the interface between the two committees opened the way to an evolving learning process that was more likely to produce a final result that would be acceptable to all Member States. When all is said and done, the representatives of the governments of Member States are those who will ultimately accept or reject any normative or standard-setting instruments.

The UDBHR (2005) was preceded by two other declarations, *The Universal Declaration on the Human Genome and Human Rights* (1997) and *The International Declaration on Human Genetic Data* (1993). In 2001, the General Conference adopted a resolution inviting the Director-General to submit the technical and legal studies regarding the possibility of elaborating universal norms on Bioethics (UNESCO 2001). The feasibility study prepared by the IBC concluded that it would be possible to prepare a Declaration and this was accepted by the Executive Committee. Then the IBC, composed of experts on the issues in question, was asked to draft a text which would be revised by government experts and then submitted to the General Conference that would approve the text, ask for more work to be done or reject it.

Member States asked the IBC to conduct extensive consultations with them, and with the scientific community and civil society. Those consultations were conducted in a range of countries (including Russia, Turkey, Iran, Lithuania, the Philippines and Mexico) and the different versions were discussed with governments, researchers and civil society representatives, in order to better allow for the different contexts in which this Declaration would apply. Extraordinary sessions and online consultations were also held. So, even though some critics continued to maintain that the consultation process had not been broad enough, it can be reasonably argued that large-scale consultation did take place.

This bureaucratic and politically charged process is often difficult to understand and frustrating for committees of experts. But if, at the end of the process, the proposed instrument is adopted and the text remains basically similar to the one drafted by the committee of experts, the outcome can be deemed successful because, after all, UNESCO is not a university, a regional organization or an association, it is an international forum of decision makers.

Discussions between the IBC, the IGBC and intergovernmental experts during the drafting process for the definition and scope of the Declaration: whether or not to move away from a medically centered conception of Bioethics

It is informative to provide an overview of certain discussions between the IBC, the IGBC and the intergovernmental experts, especially as regards conceptions of Bioethics and the follow-up actions that were adopted at the 2005 General Conference.

In the past decades, Bioethics has moved from a medically centered discipline to a universal and socially focused discipline that must and should concern all citizens. Today, it corresponds more closely to the all-embracing views of Van Rensselaer Potter, whose definition encompasses peace, poverty, ecology, human well-being and the survival of the human race, than it does to the individually and medically focused approach of André Hellegers. In this regard, it is interesting to reflect on how positions evolved during the finalization of the UDBHR; the discussions are a telling reflection of the different conceptions of Bioethics.

During the elaboration of the Declaration, the IBC, the IGBC and government experts had long, in-depth discussions about the definition of Bioethics and the scope of the Declaration. Some countries wanted to keep a medically centered approach while others wanted to introduce a broad-reaching definition of the concept so as to include the social sciences. Ultimately, IBC members proposed the following wording for the final text of the Declaration:

Article 1 – use of terms

For the purpose of this Declaration:

(i) The term 'bioethics' refers to the systematic, pluralistic and interdisciplinary study and resolution of ethical issues raised by medicine, life and social sciences as applied to human beings and their relationship with the biosphere, including issues related to the availability and accessibility of scientific and technological developments and their applications;
(ii) The term 'bioethical issues' refers to the issues mentioned in Article 1 (i) and
(iii) The term 'decision or practice' refers to a decision or practice arising within the scope of this Declaration and involving bioethical issues.

Article 2 – scope

The principles set out in this Declaration apply as appropriate and relevant:

(i) To decisions or practices made or carried out in the application of medicine, life and social sciences to individuals, professional groups, public or private institutions, corporations or states;
(ii) To those who make such decisions or carry out such practices, whether they are individuals, professional groups, public or private institutions, corporations or states.

(IBC 2005)

The discussions that took place during the two meetings of Intergovernmental Experts in April and June 2005 around Article 1 clearly reflect the differing points of view.

In the final text (see below), the absence of a definition and the wording of the scope are clear indications of the desire of the Member States to exercise caution as regards the adoption of too broad a conceptual approach to Bioethics. Social sciences were not included per se in the wording of the scope; the compromise was to say that the Declaration applied to human beings "taking into account their social, legal and environmental dimensions".

The Report of that first meeting of Intergovernmental Experts says:

> The discussion dealt first of all with the scope of the Declaration. Some participants said that the field of application of bioethics had been considerably extended in recent years; these participants also said that, although bioethics originally referred to ethical issues arising in the field of medicine and life sciences, over the past ten years it had gradually encompassed ethical issues associated with the environment and biosphere; it had acquired a particularly strong social dimension, notably in developing countries. Whilst this present broad field of application of bioethics was not questioned, divergence appeared with regard to the nature of bioethical issues that should fall within the field of application of the Declaration. Some wished to limit the scope of the Declaration to bioethical issues related to medicine and the life sciences, at the same time expressing the wish that the text acknowledge the link between the human being and the biosphere. Others felt that the social dimension of bioethics should be at the heart of the future Declaration, the principles of which should apply not only to so-called "emerging issues", i.e. those linked to advances in science and the new technologies, but also to "persistent" issues, i.e. those linked to development, poverty, public health, access to treatment and health care, etc.
>
> (UNESCO 2005a)

The final proposed compromise text agreed upon during these meetings of Intergovernmental Experts and adopted by Member States at the General Conference has no definition of Bioethics but starts with the scope and states:

Article 1

1. This Declaration addresses ethical issues related to medicine, life sciences and associated technologies, as applied to human beings, taking into account their social, legal and environmental dimensions.

2. This Declaration is addressed to states. As appropriate and relevant, it also provides guidance to decisions or practices of individuals, groups, communities, institutions and corporations, public and private.

(UDBHR 2005)

Member States were also careful to protect their power over other groups in society by starting paragraph 2 of the scope by saying that the Declaration is addressed to states, thus ensuring that they would retain the right to be the first to decide, while the IBC had identified states as one among several stakeholders in society.

The IBC had been future-oriented in its proposed text and more in line with a broad, global definition of Bioethics, with a mode of governance that included an open and transparent model of decision-making. Member States, though not entirely resistant to these concepts, were hesitant to engage as far as the IBC was proposing. It is probably fair to say that the final text is a good illustration of a pragmatic consensus.

It is also important to note that the recommendation of the IBC to change the title of the Declaration from *Declaration on Universal Norms on Bioethics* to *Universal Declaration on Bioethics and Human Rights*, was accepted, based on the fact that the concept of universal norms was not acceptable to many countries. On the contrary, a 'Universal Declaration' with principles that are interrelated and complementary and can be "considered in the context of the other principles, as appropriate and relevant in the circumstances" (Article 26) provides the conditions that enable the diversity of cultural contexts to be taken into account.

Implementation and follow-up action by UNESCO

The first two paragraphs, a) and b), of the text proposed by the IBC in Article 24 on The Role of States were accepted by Member States and became Article 22 in the final Declaration.

The text proposed by the IBC and subsequently accepted reads as follows:

Article 24- role of states

a) States should take all appropriate measures, whether of a legislative, administrative or other character, to give effect to the principles set out in this Declaration, in accordance with international human rights law. Such measures should be supported by action in the spheres of education, training and public information. States should also take appropriate measures to involve young people in these activities.
b) States should encourage the establishment of independent, multidisciplinary and pluralist ethics committees, in accordance with Article 20.

(IBC 2005)

Article 24 clearly defines the actions that should be taken by Member States following the adoption of the Declaration. Furthermore, there was in the IBC's proposed text an Article 27 entitled: The Roles of the International Bioethics Committee (IBC) and the Intergovernmental Bioethics Committee (IGBC) where it was specified that the two committees "shall contribute to the implementation of this Declaration and the dissemination of the principles set out herein".

There was also a paragraph in this same article requiring Member States to report every five years on the steps they had taken to implement the Declaration. This article was later removed by the Government Experts.

The IBC had also proposed an article on Follow-up action by UNESCO that clearly outlined the evaluation process:

Article 28- follow-up action by UNESCO

UNESCO shall take appropriate action to follow up this Declaration by evaluating new developments in science and technology and their applications according to the principles set out herein.

a)   UNESCO shall reaffirm its commitment to dealing with the ethical aspects of the biosphere and, if necessary, shall endeavour to elaborate guidelines and international instruments as appropriate, on ethical principles related to the environment and other living organisms.

b)   Five years after its adoption and thereafter on a periodical basis, UNESCO shall take appropriate measures to examine the Declaration in the light of scientific and technological developments and, if necessary, to ensure its revision, in accordance with UNESCO's statutory procedures.

c)   With respect to the principles set forth herein, this Declaration may be further developed through international instruments adopted by the General Conference of UNESCO, in accordance with UNESCO's statutory procedures.

(IBC 2005)

The follow-up responsibilities described in the final text adopted by the Committee of Intergovernmental Experts and by the General Conference are far less onerous:

Article 25- follow-up action by UNESCO

a)   UNESCO shall promote and disseminate the principles set out in this Declaration. In doing so UNESCO should seek the help and assistance of the Intergovernmental Bioethics Committee (IGBC) and the International Bioethics Committee (IBC);

b)   UNESCO shall reaffirm its commitment to dealing with bioethics and to promoting collaboration between IGBC and IBC.

(UDBHR 2005)

It is clear that Member States wanted to avoid making too strong a commitment on what they would need to do in the future and what they should be prepared to explain and report on, as this would make overly heavy demands on them.

At the same time, it is becoming increasingly obvious that more modern processes of reporting need to be developed to monitor and evaluate actions taken by Member States. Some effort has been made in the past few years to move in that direction.

In addition, the need to address weaknesses in follow-up action has not prevented the Secretariat and many Member States from working to build up their own strengths in Bioethics.

Moving from adoption to action

Even though the follow-up process adopted is not as strong and explicit as the IBC would have liked it to be, much has been done to keep the Declaration alive and to develop tools and mechanisms for its implementation, in other words, to move from paper commitments to effective action (Stanton-Jean 2016).

As of 2013, Ethics Committees have been put in place and supported in more than 17 countries. In addition, many other countries in different parts of the world have approached UNESCO for assistance in creating similar structures (UNESCO 2013). On the implementation side, training material like the Core Curriculum (UNESCO 2008), Study Materials (UNESCO 2011) and different casebooks, as well as the booklet guide on setting up Bioethics Committees, are credible support tools for an effective implementation process. Training courses have been conducted in many countries and 12 UNESCO Chairs have been established. A global ethics observatory (GEOBS) has been set up to inform countries and organizations about programs, expert resources and legislation that can serve as examples or references. The IBC has developed material to explain the principles of the Declaration and has produced reports done by its experts on a variety of related topics.

The widespread impact of the Declaration was manifested when UNESCO's financial situation had to be reviewed following budget cuts. A grid was developed to identify which programs should receive more funding than others, and Bioethics came very high on the final tables produced for the Social Sciences.

The standard-setting work of UNESCO has contributed over the years to knowledge-building in many fields. In Bioethics the challenge for the future is to strengthen the global research agenda in this field. As Professor Abi-Saab has observed, these instruments:

> helped transform, through an incremental and cumulative process over some three or four decades, what were initially vaguely perceived as remote and abstract propositions about the common interests and values (i.e. the public goods) of a hardly discernible international community, into current and palpable concepts, familiar to large sectors of international public opinion.
>
> (Abi-Saab 2007)

The observation is particularly true in the case of Bioethics. The elaboration and implementation of the UDBHR was a difficult process. But there is good reason to contend that Member States showed courage in agreeing to adopt and work with this instrument.

Conclusion

The UDBHR has now been in existence for 12 years and the time has come to set an action and research agenda for the future. This agenda needs

> a program of work of the IBC and the IGBC that will be forward looking. The challenge will be to deal with social, scientific and cultural issues that are already facing us like, among others, cultural diversity, power sharing between scientists, governments, civil society and corporations and global disasters (such as infectious diseases), end of life issues, new technologies, academic researches (sic) on the definition of global Bioethics as a discipline and a praxis.
>
> (Stanton-Jean 2016)

The UDBHR, with the participation of countries from all the different regions of the world, and especially with the important contribution of strong Member States like Brazil, South Africa, China, India and others, has led to the development of major principles such as social responsibility, benefit-sharing, the protection of future generations, solidarity and cooperation, equality, justice and equity, as well as respect for human vulnerability, all values that are essential to the development of global Bioethics because: "Global problems can no longer be approached only from an exclusively Western (or Eastern or Southern) perspective; rather they require a really global perspective" (ten Have 2016).

References

Abi-Saab, Georges 2007, *Introduction, in UNESCO: Standard-Setting in UNESCO. Vol. 1, Normative Action in Education, Science and Culture, 219*, ed. Abdulqawi A. Yusuf, Paris: UNESCO Publishing.

Global Ethics Observatory (GEobs), <www.unesco.org/new/en/social-and-human-sciences/themes/global-ethics-observatory/> last visited 26-05-2017.

IBC 2005, *Preliminary Draft Declaration on Universal Norms on Bioethics, Recommended title: Universal Declaration on Bioethics and Human Rights*, Finalized on 09-02-2005. Ref. SHS/EST/CIB-EXTR/05/CONF.202/2.

Stanton-Jean, M. Reznik 2016, The IBC Universal Declarations: Paperwork or Added Value to the International Conversation on Bioethics? The Example of the Universal Declaration on Bioethics and Human Rights, in *Global Bioethics: The Impact of the UNESCO International Bioethics Committee* (pp. 13–21) Bagheri, A., Moreno, D., Semplici, S., eds., Dordrecht: Springer Publisher.

ten Have, Henk 2016, Globalizing Bioethics Through, Beyond and Despite Governments, in *Global Bioethics: The Impact of the UNESCO International Bioethics Committee* (pp. 1–11) Bagheri, A., Moreno, D., Semplici S., eds., Dordrecht: Springer Publisher.

UDBHR 2005, *Universal Declaration on Bioethics and Human Rights*, <http://www.unesco.org/new/en/social-and-human-sciences/themes/bioethics/bioethics-and-human-rights/> last visited 12-07-2017.

UNESCO 1945, *Constitution*, <http://portal.unesco.org/en/ev.php-URL_ID=15244&URL_DO=DO_TOPIC&URL_SECTION=201.html> last visited 12-07-2017.

UNESCO 2001, *Resolution 22*, 31st Session of the General Conference of UNESCO, <http://unesdoc.unesco.org/images/0012/001246/124687e.pdf> last visited 12-07-2017. Also referenced in: ten Have, Henk, Michèle, S. Jean (eds.) 2009, *Universal Declaration on Bioethics and Human Rights: Background, principles and application*, Paris: UNESCO Publishing.

UNESCO 2005a, *Report of the First Intergovernmental Meeting of Experts Aimed at Finalizing a Draft Declaration on Universal Norms on Bioethics: UNESCO Headquarters*, Paris 4–6 April, 2005. Ref. SHS/EST/05/CONF.203/5.

UNESCO 2007a, *Standard-setting in UNESCO, Vol. 1: Normative Action in Education, Science and Culture, 12*, ed. Abdulqawi, A. Yusuf, Paris: UNESCO Publishing.

UNESCO 2007b, *Standard-Setting in UNESCO, Vol. 1: Normative Action in Education, Science and Culture, 13*, ed. Abdulqawi, A. Yusuf, Paris: UNESCO Publishing.

UNESCO 2008, *Bioethics Core Curriculum, Section 1: Syllabus, Ethics Education Program, Paris*, <http://unesdoc.unesco.org/images/0016/001636/163613e.pdf> last visited 12-05-2017.

UNESCO 2011, *Study Materials, Section 2: Ethics Education Program*, Paris, <http://unesdoc.unesco.org/images/0021/002109/210933e.pdf> last visited 25-05-2017.

UNESCO 2013, *1993–2013: 20 Years of Bioethics at UNESCO*, last visited 25-05-2017.

III The Universal Declaration on Bioethics and Human Rights as a landmark in the development of global bioethics

Henk ten Have

Introduction

Philosophy does not start with wonder but with indignation. This statement of philosopher Simon Critchley is perhaps even more pertinent for ethics. In a world where many people are richer than ever before, where we have more sophisticated science and technology to improve human well-being, and where we have more and better communication facilities than our ancestors, we are often disappointed, and sometimes exasperated and horrified by problems of migration, inequality, war, terrorism, violence, and climate change (Critchley 2012). Similar experience of failure and injustice initiated the emergence of bioethics in the 1970s. Traditional medical ethics was transformed because of disturbing experiences such as scandals of medical research and challenges of technological interventions that could bring great benefits but also serious harm and impersonal, dehumanizing care. The rise and expansion of global bioethics since the turn of the millennium is stimulated by disturbing cases such as female genital cutting, people dying from treatable diseases, the use of different standards of care in research in developing countries, women exploited as commercial wombs, the Trovan case in Nigeria, and mismanagement of Ebola virus disease in West Africa. We are continuously confronted with a world that is thoroughly unjust and exploitative.

The moral indignation and dissatisfaction that befalls us motivates action because the ethical subject faces not abstract notions like justice or solidarity but concrete demands of other people, not in general but as particular human beings: strangers, marginalized, vulnerable, and excluded. This is one of the impacts of globalization: the misery and misfortunes of individual strangers invades our private existence. But ethics at the same time presents universal claims. Because of our shared interconnectedness and vulnerability, we are committed to the demand of the other. This is the same point made by French philosopher Alain Badiou: confronted with inhumanity, we encounter a universal address that makes us search for new possibilities within a particular context. Universality is situated. The singular is always related to the universal. This is not just ethics; it is simultaneously politics; in the words of Badiou: politics is "the local creation of something generic" (Badiou 2015: 56). The connection between ethics and politics is also articulated by Critchley (2012: 132): Politics is "an ethical practice that is driven by a response to situated injustices and wrongs".

In this perspective, global bioethics is not the imposition of a value system. It is engagement in a dialectical and intercultural process of interaction between global principles and local practices, a contentious intercourse between 'above' and 'below'. It is also a continuous interplay of theoretical discourse and practical implementation. Global bioethics therefore is not a ready-made product but in process. It is the aspiration to realize the universal in the local. But it is first of all a social ethics that goes beyond the view that ethics is primarily a matter of personal commitment and individual lifestyle. Global bioethics presents a horizon of reflection, analysis, and action that brings ethical principles associated with commons, cooperation, future generations, justice, protection of the environment, solidarity, social responsibility, and vulnerability (back) into the debate of globalization.

This chapter will examine the emergence of global bioethics, and specifically focus on the role of the Universal Declaration on Bioethics and Human Rights. It will first discuss the need for a global bioethical approach that differs from mainstream bioethics that has evolved since the 1970s.

The need for global bioethics

When Van Rensselaer Potter coined the term 'global bioethics', he wanted to salvage his initial conception of bioethics that he introduced in 1970 (Potter 1988). He observed that the neologism was quickly used and disseminated but in a different way as he had intended. It was in fact a new label for usual business, even if it covered more issues and included a wider range of actors. For Potter 'bioethics' (or 'biomedical ethics' or 'healthcare ethics', as it was also named) was not much different from the traditional medical ethics but now applied by a new type of professional (most often philosophers and theologians rather than healthcare professionals). It still was primarily focused on medical issues and on ethical questions arising in the individual relationship between patients and healthcare providers. His conception of bioethics intended to broaden the scope of ethical discourse. In his view, ethics should go beyond the individual level of analysis and should focus on social and environmental concerns. The major ethical challenges of today are related to overpopulation, war and violence, poverty, and environmental pollution. These problems can no longer be addressed with the usual ethical theories but require new approaches and intensive interdisciplinary cooperation between scientists and moral experts. 'Global bioethics' in Potter's vision is an appropriate word to indicate that a broader and more encompassing effort is needed. It also clarifies that many problems today are worldwide and no longer confined to the borders of countries and cultures.

Potter's vision, though long-time ignored, is nowadays recognized in the emerging area of global bioethics. It is obvious that many moral problems of today are global in nature; they affect all people wherever they are residing. Problems that were localized in the past are now crossing borders, such as insecurity, environmental degradation, and water scarcity. Moral issues that are regulated within borders become difficult to control because people and technologies are mobile. Practices that are outlawed in some countries are offered for sale in other countries. There are growing illegal markets for organ trade, human trafficking, and drugs. The global nature of moral problems affecting health and healthcare implies that they cannot be addressed at local or regional levels but require global governance. It furthermore implies that a global dialogue is necessary. Since present-day problems can affect anyone everywhere, all people should be involved in how they are solved or at least mitigated. Solutions can no longer be based on technologies, ideas, and approaches of the global North but should engage people and populations across the globe. Another implication of the global nature of bioethical challenges is that it highlights the global nature of bioethics itself. Although ethics always has a universal core, exemplified in the Golden Rule or the 'moral point of view', demanding to go beyond individual interests, the processes of globalization demonstrate that the circle of moral concern also in practice has been expanding. Human history shows the possibility of moral progress. This history can be written as a long concatenation of cruelties, barbarisms, and terrorisms, showing that ethical discourse is impotent and futile. But detailed studies argue that today we are living in the most peaceable time of human existence. Steven Pinker (2011) for example argues that violence in all its forms (such as murder, rape, torture, civil wars, genocide, and terrorism) has substantially declined. Explanations for these reductions may differ, but it is clear that they are associated with important changes in moral sensibilities and behaviors: increasing self-control, the emergence of a culture of dignity and cooperation, growing empathy with fellow human beings. Over time, the moral horizon seems to have enlarged so that the perspectives of other persons (slaves, women, children, racial minorities, and homosexuals) are indeed taken into account. Kenan Malik (2014) explains that the idea of a universal community is not the privilege of a specific culture or civilization; it has developed in various civilizations. The concept of universalism is not the product of Western imperialism.

Gradually, the circle of moral concerns has expanded, so that one could abstract from direct and immediate experience and develop sympathy for more groups of people. This has not only impacted violence but also long-standing phenomena like slavery, interstate wars, spousal abuse, infanticide, and child abuse that have become intolerable. Moral progress has been accomplished because the circle of ethics has expanded in the course of history: from family and tribe to nation and state, and to all human beings, and perhaps to animals and nature. The theory that the circle of ethics has expanded is especially promoted by philosopher Peter Singer (2011). But this expansion of ethical practice also reflects the development of ethics predicted by Potter's colleague Aldo Leopold, moving from a focus on individuals, to society, and to the environment, with bioethics, at least in the vision of Potter, as the final stage (ten Have 2012).

The contribution of the Universal Declaration on Bioethics and Human Rights

This background of globalization has two consequences for bioethics. First is the specific nature of contemporary bioethical problems. They demonstrate that individual, society, and environment are interconnected. A broad and encompassing ethical approach is necessary to provide the moral resources to address these problems. This is one of the reasons why global bioethics is indispensable today. It also clarifies why the Universal Declaration on Bioethics and Human Rights is a landmark in global bioethics. It provides an ethical framework for analyzing and criticizing current challenges. The second consequence is that bioethics is not a finished product that can be imported and applied regardless of local circumstances. Global and local approaches are not contrasted but interacting, and thus influencing each other. Since bioethics itself is part of culture, it cannot be imposed. It will be transformed in a dialectical process of interaction between global and local approaches. Interaction and exchange mean adaptation and modification, and sometimes rejection. In this process of cultural exchange in many different settings, a consensus will gradually emerge on a comprehensive approach. Globalization implies accommodation rather than assimilation. Global responsibilities and universal values will be articulated that for the time being are not universally accepted but that are applicable to all human beings wherever they are, because they are justifiable on the basis of reason and common interests. Such bioethical framework will be the result of deliberation and negotiation. Ethical approaches will converge towards commonly shared values over time, finding a balance between local interests and global obligations, between partiality and impartiality (Veatch 2012). It is precisely this dialectics between global and local that has led to the adoption of the Universal Declaration on Bioethics and Human Rights. Although many bioethical issues are not settled, the text is very general and not legally binding, mechanisms of implementation are weak, and reporting and monitoring procedures missing, it provides an ethical framework to address global issues. This not merely offers common perspectives but, more important, an intercultural space in which to recognize divergence of moral views and to engage at the same time in a continuous search for common principles. A global world is not simply characterized by the existence of multiple value systems, but first of all by the interaction and reciprocal learning of these systems. 'Interculturality' emphasizes interaction. 'Inter' refers to separation but also linkage and communication. It acknowledges diversity while insisting on global values and common perspectives. Interculturality introduces a moral vocabulary of interaction, dialogue, participation, trust, cooperation, and solidarity. It articulates an interest in what unites people rather than what divides them. Intercultural dialogue is driven by the quest for unity or rather commonality, not uniformity. Convergence will only result from a persistent process of 'translation'. Since there is not one moral language that is pre-given, underlying, or more fundamental than the other languages, there is no possibility to step outside of the communication process. A 'trans-cultural' reference point that is universally acceptable does not exist. There are only 'interstitial' spaces where cultures interact and overlap, and where people communicate. The first step in reaching understanding is to recognize that there are radically different 'languages'. Consensus only becomes a possibility after differences are expressed and acknowledged. We have no choice other than to interact in our various languages and search for common understanding. Convergence, as demonstrated in the Universal Declaration on Bioethics and Human Rights, is not given but the result of an ongoing activity of deliberation, consultation, and negotiation.

A broader framework for bioethics

The adoption of the Universal Declaration can be considered as the result of mainstream bioethics as it has evolved since the 1970s in most developed countries. However, it is a mixed result, partly due to the success and popularity of bioethics, and partly due to its limitations. Mainstream bioethics was quickly consolidated with the establishment of institutes, centers, national committees, teaching programs, journals, policy guidelines, and special legislation. In a relatively short time bioethics, initially conceived as a vision, became its own separate discipline, regarded as a branch of applied ethics. The new discipline is governed by the paradigm of principlism. Methodologies and practical approaches are based on the principles of respect for autonomy, beneficence, non-maleficence, and justice. Principlism is attractive and effective. Principles provide a source of moral knowledge that is open to reason and experience. They are commonly shared and accessible for everyone with reason and experience. Principlism has the same advantages as Jacques Maritain attributed to human rights: we agree to have them and specify what they are without agreeing why we have them (Ignatieff 2012). This paradigm furthermore illustrates the universal aspirations of ethics. It formulates a view from nowhere in emphasizing what human beings have in common against the view from somewhere, i.e., the particularities of family, community, ethnicity, religion, and nation in which they differ. This is exactly the function of ethics, according to Ignatieff: it calls the particular to justify itself before the universal, and subject it to a demand of justification. In his words: "The essence of moral life is this process of recurrent, repeated, behavior-changing justification. This process needs standards – a global ethics provides the view from nowhere, global ethics provides a view from somewhere" (Ignatieff 2012: 18).

However, the problem is that despite its claims, mainstream bioethics is not really global bioethics. In interrogating particularism it does not reflect on its own particular origin and setting. Mainstream bioethics is not a view from nowhere; it has emerged within a specific cultural context that articulates the primacy of individual autonomy. It also proceeds from a particular view of ethics as a continuous process of justification, while moral life is more than reasoning and justification. For most of its history, ethics in the West itself has not only been a doctrine, discourse, or theoretical activity but a way of life, an existential choice, an exercise in practical ethics, and concern about the world, not unlike in many other philosophical and spiritual traditions in other parts of the world (Hadot 1995). Therefore, ethics is not simply the application of principles to complex situations. It is not merely a theory to consider in practice. It is more than the effort to distinguish right from wrong, good from bad. Ethics also is a quest to understand what it means to be human. It involves moral experience, moral sensitivity, social concerns, and public virtues.

In light of these limitations of mainstream bioethics, a broader ethical framework is required that not only proceeds with a global set of principles but also an approach that goes beyond the application of principles to various practices. The Universal Declaration on Bioethics and Human Rights (UDBHR) does exactly that: it presents a wide range of relevant principles for addressing global issues as well as a view of ethics that recognizes the tension between principles and practices, and the impossibility to impose global principles in different social, cultural, and political contexts.

A global ethical framework

In Potter's vision contemporary bioethics should be oriented towards the future; a bridge between the present and the future because for the survival of humankind it is vital to focus on long-term interests and goals. The ultimate goal of bioethics is long-term global human survival. For this purpose, new knowledge and old wisdom should be combined. Otherwise, the pressing global problems of today cannot be properly addressed. Another point is that the goal of human survival can only be accomplished when there is a balanced social and ecological system for human life to flourish. Bioethics therefore should do more than presenting an ethical framework focused on the individual good; it should also refer to the common good as an anchor for social justice and environmental sustainability.

The ethical principles declared in the UDBHR provide precisely such an expanded framework.

The principles determine the different obligations and responsibilities of the moral subject (*moral agent*) in relation to different categories of moral objects (*moral patients*). The principles are arranged according to a gradual widening of the range of moral objects: the individual human being itself (human dignity; benefit and harm; autonomy), other human beings (consent; privacy; equality), human communities (respect for cultural diversity), humankind as a whole (solidarity; social responsibility; sharing of benefits) and all living beings and their environment (protecting future generations and protection of the environment, the biosphere, and biodiversity). Although no hierarchy among principles is implied, this ordering means that the focus of ethical concern is not restricted to an individual point of view; ethical concern should equally take into account common social and ecological perspectives. Global bioethics therefore cannot be merely medical, social, or environmental; it should include all perspectives at the same time. At a global level, bioethics should have a wider focus and be concerned with humanity as a whole.

A different view of implementation

Bioethical problems are not just theoretical queries. They have practical implications. Confronted with various options for treatment or intervention, patients and health professionals have to decide what is best to do. Various values are at stake and they cannot all be respected. In mainstream bioethics this is often the type of ethical challenge that is at the table. It also reflects the paradigmatic setting for ethics; individual persons must decide what they ought to do when facing situations in which values conflict with each other. The choice made should be justified with rational arguments and should be based on their own values. The fundamental ethical question is: What should I do? To answer this question, one should take 'the moral point of view'. In the Western tradition, this means that one should take care that a decision or action may universally apply to everyone.

However, global bioethical problems have a different character (ten Have 2014). Is it ethically justified that poor people in developing countries are selling their kidneys? Should women in Vietnam be allowed to outsource their wombs to couples in Australia? Is it ethically acceptable that 1,5 million people die each year from tuberculosis while this disease is almost completely curable? Should global trade be more important than global health? These questions indicate that global problems request collective answers: What should *we* do?

Being 'global' means that a bioethical problem is affecting in principle all human beings wherever they live. Of course, problems challenge individuals and they have to determine what they ought to do in response. Many people work as volunteers to provide aid to victims of disasters; they donate money to NGOs such as Doctors beyond Borders, or work for them some time in difficult circumstances. Infertile couples are aware that commercial surrogate motherhood is available, and they have to decide if they want to use these services when they desire children. It is evident that global problems can be translated into individual ethical problems. But as typically global phenomena they do not first of all present ethical challenges at an individual level. The current concerns about the Zika virus pandemic illustrate the difference in perspective. This tropical virus, transmitted by mosquitoes, is rapidly spreading through the Americas and starting to spread in Europe. For most people the disease is like a simple flu. However, the virus is obscure. There is not much knowledge. Vaccines, treatment, and diagnostic tests do not exist. But the disease is associated with usually rare conditions that now have become more frequent, such as shrunken brains of newborns. It is probable that the disease will have a wide impact. People are advised not to travel to a growing number of countries. It seems that there is no way to stop the disease, not even with the military. The main question is what can we do and should we do? From the point of view of mainstream bioethics, focused on individual ethics, one of the main questions is whether infected pregnant women should have easier access to abortion. From the perspective of global bioethics, one of the major questions is why are there no vaccines and treatments? This is not accidental. It illustrates the 10/90 gap in health research that was already criticized in 2001. Only 10 per cent of research in global health concerns conditions that account for 90 per cent of the global health burden (Stevens 2004). An interest to address 'tropical' diseases is simply lacking. They are commercially not attractive. There are no incentives to produce vaccines since the majority of people affected are poor, and strong governments are lacking. An adequate public health infrastructure is absent; it has been dismantled through decades of neoliberal policies and privatization of healthcare, especially in developing countries. Past global policies have weakened health systems across the world.

The irony is that we have been confronted with a similar story over the past three years in connection to the Ebola epidemic in West Africa, highlighting mismanagement and lack of solidarity in the international community. The main concern of Western countries was to prevent the virus to spread, and to protect their own vulnerability, not the vulnerable people in Africa and the complete lack of health infrastructure and health professionals to cope with the disaster in that part of the world. The debate especially in the mainstream bioethics community was about potential treatments and new vaccines, and under what conditions these potential interventions could be expedited. Bioethics simply chose to ignore the social, political, and economic context in which the pandemic could emerge.

One conclusion from the global nature of contemporary bioethics problems is that they force us to go beyond an individually focused ethics. Global ethics needs a wider ethical framework, as advocated by Potter. The final reference is humanity, not the autonomous individual. That requires an ethical point of view that is shared by as many others as possible, and at least aspires to apply universally to all people across the world, assuming that in practice full agreement will never exist. Global problems cannot be addressed by individuals but ultimately require cooperation and solidarity. Respect for diversity is one of the ethical principles to take into account so that diverging moral views will continue to exist, while on the other hand global challenges such as Ebola and Zika pandemics require agreement on basic ethical principles. The need for global governance and emphasis on common perspectives may make global bioethics different from ethics in general.

Another conclusion is that the social-ethical nature of global bioethics initiates a continuous search for common perspectives, values that are shared among many cultures and religions.

This conclusion has two implications. One is that global bioethics necessarily is work in progress. It will never be a finished package of ethical principles or practice guidelines (such as articulated by 'principlism' in mainstream bioethics) that can be applied all over the globe. Global bioethics has to recognize diversity of moral views, but at the same time there will be convergence, as Veatch (2012) has argued, towards commonly shared values. For this to happen, a moral vocabulary of interaction, dialogue, participation, and cooperation is unavoidable. A common ground needs to be cultivated through interaction and communication. Convergence is not given but the result of ongoing interactive activity.

A second implication is that the view that global bioethics operates at two levels is insufficient. This view generally underlies discourses of implementation. It is argued that on the one hand, there is a set of global principles on which traditions and cultures agree; this is expressed in international human rights language and elaborated into specific bioethical principles. On the other hand, there are many efforts to articulate more specific bioethics standards within the context of particular religious and cultural settings. Global principles are always applied within local and specific cultures and traditions, and this necessitates continuous interpretation and articulation of global standards (Sullivan and Kymlicka 2007; Held 2010).

This distinction, however, is not appropriate. First, distinguishing two 'levels' suggests a hierarchy, while in fact global and local interact at the same level. Second, the universality of principles identified at the global 'level' is the outcome of interactions with the local 'level', so that the global principles are in fact shaped by particular settings and approaches. Global bioethics has several constitutive components, rather than two levels. It covers a global domain as well as local domains but also a process of continuous dialectic interaction with mutual exchange, inspiration, aspiration, articulation, deliberation, learning, and negotiation. Global bioethics is characterized by duality. It is global in the sense that it assumes a universal ethical framework (as the result of intercultural dialogue and consensus) and at the same time local in the sense that this framework necessarily has to be applied in diverse multicultural settings. It is furthermore characterized by labor. It is not simply academic reflection but work in practice.

This last implication is important. Globalization is often regarded as 'colonialism'; the imposition of a specific, often Western, ethical framework upon the rest of the world. The same is true for global bioethics: it is also blamed as imperialistic and colonizing (Qiu 2002). This vision, however, is incorrect. It assumes that globalization and localization are two different processes. In a dialectical view, global and local intermingle and interact. That means that global principles only come to life when applied in local settings; they will thus be transformed while local experiences will feed back into the global principles. In fact, an inversion of the statement of Ignatieff is at work: the universal is also called to justify itself before the particular.

This dialectic interaction between global and local means that global bioethics is not imposed to the rest of the world but transformed. Bioethics is not a product that can be imported and used regardless of the local culture. Because it is itself part of culture, exchange means adaptation and modification, and sometimes rejection. In this process of cultural exchange in many different settings, a consensus will gradually emerge on a comprehensive approach through scrutiny, analysis, debate, application, modification, and reinterpretation. It is exactly through such processes of exchange and negotiation that the Universal Declaration on Bioethics and Human Rights has emerged in the first place, as the beginning, not the closure, of a long process furthering the expansion of the circle of moral concern. The same process is at work during the implementation of the UDBHR. The lesson from current human rights scholarship is that applying global principles is 'globalization from below' rather than from above (Koh 1999; Risse, Ropp and Sikkink 2013; Hafner-Burton 2013). Rather than compliance with global principles, practical applications entail 'domestication'. Principles need to be transformed and internalized into domestic systems and local contexts. Usually, this is not done by governments but by non-governmental organizations and cooperating individuals. The challenge to translate the global principles into local settings and practice therefore is with the bioethics, scientific, and professional communities.

Conclusion

Global bioethical problems require global answers. To provide answers we need a global bioethics. Potter has proposed a concept of global bioethics that encompasses individual, social, and environmental perspectives focused on addressing global problems. The UDBHR has elaborated this concept into an articulated ethical framework with a broad range of ethical principles. The adoption of the UDBHR was the result of deliberation and negotiation based on a dialectic process of interaction between global and particular moral views. The outcome is provisional since the dialectic nature of global bioethics will not deliver a closed set of ethical principles. More important than the principles will be their application in practical settings leading to their continuous debate, modification, and transformation. The same dialectic process that has given rise to the Declaration will be required in its implementation. This is a task of the governments that have unanimously adopted this document, but much more so of actors in countries, regions, and contexts that are faced with the same global bioethical challenges as all of us.

References

Badiou, Alain 2015, *Philosophy for Militants*, London and New York: Verso.

Critchley, Simon 2012, *Infinitely Demanding: Ethics of Commitments, Politics of Resistance*, London and New York: Verso.

Hadot, Pierre 1995, *Qu'est-ce que la philosophie antique? Gallimard*, Paris: Folio.

Hafner-Burton, Emilie M. 2013, *Making Human Rights a Reality*, Princeton and New York: Princeton University Press.

Held, David 2010, *Cosmopolitanism. Ideals and realities*, Cambridge, UK and Malden, MA: Polity Press.

Ignatieff, Michael 2012, Reimagining a Global Ethics, *Ethics & International Affairs*, 26(1): 7–19.

Koh, Harold 1999, How Is International Human Rights Law Enforced?, *Indiana Law Journal*, 74(4): 1397–1417.

Malik, Kenan 2014, *The Quest for a Moral Compass: A Global History of Ethics*, London: Atlantic Books.

Pinker, Stephen 2011, *The Better Angels of Our Nature*, London: Penguins Books.

Potter, Van Rensselaer 1988, *Global Bioethics: Building on the Leopold Legacy*, East Lansing: Michigan State University Press.

Qiu, Renzong 2002, The Tension between Biomedical Technology and Confucian Values, in *Cross-Cultural Perspectives on the (Im)possibility of Global Bioethics* (pp. 71–88) Tao, J., ed., Dordrecht, Boston and London: Kluwer Academic Publishers.

Risse, Thomas, Ropp, Stephen C., Kathryn, Sikkink (eds.) 2013, *The Persistent Power of Human Rights: From Commitment to Compliance*, Cambridge, UK: Cambridge University Press.

Singer, Peter 2011 (original 1981), *The Expanding Circle: Ethics, Evolution, and Moral Progress*, Princeton and Oxford: Princeton University Press.

Stevens, Philip 2004, *Diseases of Poverty and the 10/90 Gap*, London: International Policy Network, <http://who.int/intellectualproperty/submissions/InternationalPolicyNetwork.pdf>.

Sullivan, William M., Kymlicka, Will (eds.) 2007. *The Globalization of Ethics*, New York: Cambridge University Press.

ten Have, Henk 2012, Potter's Notion of Bioethics, *Kennedy Institute of Ethics Journal*, 22(1): 59–82.

ten Have, Henk 2014, *Global Bioethics: An Introduction*, London: Routledge.

Veatch, Robert 2012, *Hippocratic, Religious, and Secular Medical Ethics*, Washington, DC: Georgetown University Press.

IV Charters of rights and bioethical principles

A multicultural challenge

Fabio Macioce

Introduction

Bioethics is one of the several branches of the human rights tree's trunk (Annas 2004), a tree that takes roots in the tragedies of World War II and, particularly, in the Holocaust. Given this connection, it is easy to understand why the ideas of a universal framework of principles, and of an objective dimension of human dignity, have been so widespread and so successful. The concrete experience of such a tragedy, and of such an inconceivable violence, have nourished the effort to recognize to every human being, merely by virtue of his or her humanity, a same set of rights, in order to protect his or her intrinsic dignity. Both international law, and the immense corpus of treatises, as well as many other ethical declarations, can be read in that perspective.

Furthermore, human rights tradition is the core of the United Nations' action, and shows its most fundamental sense, its *raison d'être*, as it is proclaimed in its Charter, which states that they are "for all without distinction": human rights are neither generous concessions, nor privileges for any élites, and the United Nations is primarily settled and committed to promote "universal respect for, and observance of, human rights and fundamental freedoms". For that reason, not only human rights affirm a universal consensus on some ethical standards, but even the necessity – for their implementation – of international Bodies and Organisms.

In that perspective, a fundamental aspect of human rights tradition is that of being an endless effort. According to Hannah Arendt, what is at stake is "the right to have rights", that is the right to be recognized as a human being, worthy of respect and protection. Therefore, the human need for rights is endless, as much as (dramatically) the threats of our dignity. By saying that justice and natural law are *àgraphoi nòmoi* (Aristotle, *Rhetoric*, I, 10), ancient philosophers were expressing the same idea: no Charter, no list of rights, no formal international document can pretend to be the ultimate one, the one where rights can find their definitive formulation and wording, because as a matter of principle it is impossible to write them conclusively. There doesn't exist a way to express the idea of human dignity, for it is inexhaustible.

This is the reason why international guidelines or statement relating to bioethics are relatively recent. They have been the answer to the progressive awareness of the need of international standards and rules, which are aimed to protect human dignity from the threats connected to scientific and medical progresses. Among the many, the UNESCO Universal Declaration on Bioethics and Human Rights (Declaration) remarkably contributed to define common standards relating to bioethics. In other words, it has raised to the international system of human rights some principles previously considered as merely confined within the realm of medical ethics and within national bonds. Due to the authority of the Declaration, these principles have evolved from the field of national moral and political debate regarding bioethics, to the sources of international law (Faunce 2014).

The Declaration is in fact one of the first international legal Charters that expressively and comprehensively stressed the linkage between human rights and bioethics (Andorno 2007). And despite the fact that it is a non-binding instrument, as well as the majority of other similar documents, which have been issued by non-governmental organizations, the UNESCO Declaration is a highly remarkable achievement.

We are, in other words, watching the outcomes (may be partial, but notwithstanding remarkable) of a long-time route; and even if no one can say that a definitive result has been reached, we have come a long way. Twelve years (from 2005 to now) is a considerable time space, and many documents and charters have been issued from international and non-governmental organizations (Langlois 2013).

We are, therefore, in a position that legitimates, at least from a theoretical point of view, two basic questions about rights and charters, with particular reference to the field of bio-ethics: first, do we really need charters of rights, in such a pluralistic world, and facing such a rapidly evolving world, or they are a too *rigid* tool, something theoretically appreciable but practically useless? Second, are rights and principles, like so expressed in charters, suited for regulating bioethical dilemmas, or the joint pressure of biomedical evolution and cultural pluralism shows the inadequacy of any attempt to define common principles and stable guidelines?

Rights and principles in bioethics: a premise

One caveat must be stressed. On the one hand, the fact that these documents represent for many people one of the most significant achievements of our times should be stressed, whatever we think about the future of charters of rights. In other words, the opting for rely-ing on human rights, thereby setting up minimal standards in a specific field and protecting the fundamental worth of human beings, is perfectly sound with the value that our culture, wrongly or rightly, assesses to the system of international Charters (Andorno 2007). At the same time, it is not doubtful that the mechanism of human rights is one of the few, if not the only one, that can work as a framework to respond to modern challenges, by the estab-lishing of a minimal global ethics (Thomasma 2008).

On the other hand, even recognizing these merits, the international standard-setting activity can be criticised in itself. Even considering the historical merits and political value of that system, the weak contribution of such declarations (from a concrete point of view) can be pointed out. Due to their context of origin, and to political contingencies, all these charters appear to be so vague and minimalist, not providing any concrete legal tool, and leaving too wide margins of interpretation. And the joint effect of principles' vagueness and courts' interpretation makes these charters excessively ambiguous, likely to be interpreted in too different perspectives. Instead of providing shared standards to protect human dig-nity, international declarations appear to be so general and vague that they risk to become useless (Macklin 2005; Benatar 2005). This perspective is consistent with a more general argument, according to which any appeal to dignity seems to be problematic and puzzling, not giving any "clear and unambiguous guidance" in difficult or paradoxical situations, because dignity is in itself a vague and ambiguous concept (Macklin 2003).

Both these perspectives are correct. At the same time, one can say that charters of rights are *politically* appreciable and concretely useful to protect dignity for all human beings, and that rights, which are so widely recognized, are unavoidably vague and opened to many interpretations, and even to conflicting interpretations.

We can understand why these two different points of view are at the same time possible, by considering the distinction between rights and principles, or more specifically the distinction between rights as the outcomes of rules, and rights as rules in themselves.

Principles are conceived, frequently, not as reasons for action but as reasons for norms. According to Alexy, both principles and rules generate reasons for norms, in the sense that both can generate a concrete individual legal norm (i.e. a judicial decision): both are concretely applicable in specific cases. But principles are normally conceivable as prima-facie reasons for norms, while rules are definitive reasons for action: both can represent a framework to reach a legal decision, but rules can determine a particular decision directly and not needing to be weighed against other reasons, while principles always have to be weighed against others in order to be enforced in a specific case (Alexy 2002: 59–60).

Otherwise, it is possible to distinguish principles and rules by saying that principles are norms for argumentation, while rules are norms for conduct (Gross 1969: 575). Or, that principles can be enforced in different degrees and with different strength, depending on factual and legal possibilities, while rules can merely be enforced or not enforced, without other possibilities (Alexy 2002: 60).

According to many scholars, I agree with the idea that principles are highly indeterminate, general and vague, thus conceivable as "open" norms, and thus likely to be considered as constitutive (fundamental) elements of a legal system, while rules are necessarily more precise and determinate, directly representing reasons for action. As one can see, the distinction is a matter of degree, because principles are more general and more indeterminate than rules, and not a matter of quality. On the other hand, from a practical point of view, it is correct to affirm that principles can be used to express reasons for rules, while rules can express reasons for action, and ought-judgment in a particular decision.

What is at stake, here, is that fundamental rights, when they are recognized by international agreements and charters, are normally expressed as principles. And that is because principles can represent a common language among different ethical perspectives and legal systems: they represent a peculiar form of agreement, which is consistent at the same time with the aim of establishing common standards for action, and with a persistent theoretical disagreement (is something similar to what Sunstein calls an *incompletely theorized agreement*: Sunstein 1995). Principles, inasmuch consequent to political compromises, are expressed in a vague and indeterminate form, precisely because this is the way to leave wider margins to interpretation, and to make charters more flexible and suited to different situations.

It is easy to understand why principles, and charters of rights (inasmuch they are expressed through principles) are at the same time a step forward, and something highly problematic. They undoubtedly represent an achievement, because they find common standards even starting from different ethical and political positions; but at the same time they are necessarily vague and indeterminate, thus assigning to interpreters the task to give substance to them, in somehow unpredictable ways.

Is it true, conclusively, that charters of rights and declarations are too vague and indeterminate, and that it would be better not to use them in order to protect human dignity? Is it possible to affirm that international declarations of rights are too vague to be useful? My answer is twofold: yes and no, at the same time. To the first question I do not have any hesitation to reply affirmatively, for the reasons I have expressed previously. Charters of rights are really vague and indeterminate, because they are the outcome of "incompletely theorized agreements", of the attempt to establish common standards starting from different positions. At the same time, I disagree with those who affirm that vagueness is a sufficient reason to consider these principles useless or even perilous: conversely, I consider such a trait extremely useful and important, especially while we face a multicultural and pluralistic horizon. This vagueness is precisely what allows interpreters and legislators to enforce these principles, even when they have been conceived starting from a very far cultural perspective, by extending or restricting them in a way that permits an acceptable balance with local traditions and sensitiveness. And this, as I will try to demonstrate later, is one of the main problems that international documents such as the UNESCO Declaration have to face.

International charters and bioethics: merits and shortcomings

A different question is whether or not declarations of rights are suited for bioethical problems. Indeed, and more than in other fields of human activity, the reality that we have to consider when we try to define an agreement about principles and rules is continuously changing; and this is why bioethical principles can become rapidly old or inadequate in dealing with unpredicted activities, and in responding to new challenges.

In order to address that question, we should consider the fact that such documents have been perceived as a fundamental need by many specialists and practitioners, as well as by decision makers and civil society (Berlinguer and De Castro 2003). The anxiety related to the development of life science and medical technologies, with particular regard to what can directly affect human lives and human bodies, gave rise to increasing demands for rules, which could ensure respect for human dignity and fundamental freedoms.

International pressure for the definition of common standards has therefore increased together with the broadening of the fields covered by bioethics. As rightly noted (Berlinguer and De Castro 2003: 2) "The focus has since broadened considerably. In addition to issues relating to the beginning and the end of human life, bioethics covers issues raised by the donation of human organs, tissue, cells, and gametes; the scientific, epidemiological, diagnostic and therapeutic uses of genetics", and this expansion is still ongoing.

As I said, this pressure has been particularly strong at the international level. Even if states have a pivotal role in bioethical regulation (and particularly in a multicultural perspective, as I will demonstrate later), both in promoting concrete rules to protect human dignity, and in defining medical practices and standards for scientists, it has become evident that medical and scientific practices extend their reach beyond national borders: import and export of cells or genetic material, exchanges of DNA samples and of genetic data, the definition of common standards for clinical trials, the access to health care, patents and intellectual property on medicines, and many other problems, all these experiences pointed out the need for international regulations, and for common solutions which were fair for all countries and communities.

From that point of view, one can interpret the pressure for international guidelines and rules as a demand for a deeper global justice (Macklin 2009; Pogge 2001: 15): for the inequality between rich and poor countries is increasing, and even more dramatic in the field of health care, demands for international charters of rights represented a claim for common standards of justice and dignity. In other words, they represented a claim for a more just global regulation, at least in the health care field, and in any field strictly related to it such as those of food, hygiene and education. A universal instrument on bioethics is thus useful to contribute to the recognition of "the right of everyone to the enjoyment of the highest attainable standard of physical and mental health" (Article 12, International

Covenant on Economic, Social and Cultural Rights), and it can focus the global attention on vital and basic needs of poor people, and on their need to have access to life-saving treatments, to basic hygienic conditions, to not to be exploited for first world's needs, and so on.

Besides this, the need for a Declaration in the field of bioethics can be interpreted as the result of the more general binding force of the international human rights law (Smith 2005). As rightly pointed out,

> Whatever may be the privilege of nation states, multinational corporations, civil society organisations and particular individuals to ignore international human rights law, this is not a luxury open to a United Nations agency. (. . .) This is why the initiatives taken by UNESCO, on the advice of its IBC (and IGBC) have an element of the inevitable about them. In the context of United Nations agencies, and specifically UNESCO, it was impossible to continue a discourse on bioethics without paying due regard to relevant provisions of international human rights law as it affects bioethical decisions.
> (Kirby 2009b: 324; see also Sweet and Masciulli 2011)

Bioethics is even more concerning the planet as a whole, much more than simply crossing national borders: the new bioethical space, which is concerned with public discourse about principles and values, is settled on a global scale (ten Have 2013), and thus requires rules that take into account its global dimension.

At the same time, some concerns still remain: for instance, how the need for guiding principles can be balanced with the speed of biomedical evolution should be asked. Similarly, one can wonder whether principles, once stated, should be regularly revised to ensure their accordance with new challenges and new technical possibilities, or whether they should be as much as possible fixed and stable. Again, one may ask about the interaction between international Committees, Non-Governmental Organizations, international Courts, States and national Courts, Agencies, Bioethical Committees, specialists, and so on, in interpreting and enforcing these principles, as well as about the fact that many international documents in the field of bioethics already exist, and some of them are very influential, even if only the Declaration is both adopted by governments and settled on a global (and not only regional) level.

These questions are sound and difficult, and here I can only limit the focus on a few milestones, in order to answer them: my aim is not to demonstrate that the Declaration is the best way to grant human dignity facing biomedical problems, but to argue the way in which we have to understand it and its goals.

First, the Declaration (and other similar international documents) can be highly significant if interpreted as a point of reference for future debates and developments, rather than an end point: it should be interpreted and debated, developed and implemented, extended and reduced, more than what we are used to do with other fundamental charters of rights (i.e. Constitutions).

Second, the Declaration shall be read as an attempt to *produce* a shared understanding and a stronger cohesion about ethical problems and practices involving human life and health, rather than the expression of an *existing* consensus about values that we can already take as definitive and worldwide shared.

Third, it is important to pay attention to the previously mentioned distinction between principles and rules: only a few principles, and rarely, can be directly enforced in order to regulate specific and concrete problems. Mostly, they need to be specified through laws both to define when and how they are to be enforced, and to clarify to whom is the task to define specific standards and to watch over biomedical or technical practices assigned, and finally to adapt the established standards to unforeseen circumstances.

Fourth, we should consider the Declaration from a global perspective, and recognize that it can have a different impact on different countries; it can be even more useful for those countries that lack an extensive framework of bioethical rules and norms and guidelines, because they can see it as a guarantee that "the advantages and disadvantages of scientific development and technological innovation are equally and equitably shared among all nations (and) that the standards and regulations concerning bioethical issues reflect a global perspective beyond national and regional interests and concerns" (ten Have and Jean 2009).

The Declaration as a methodological expedient

I argue that it is possible to answer the question of whether or not we need, in the field of bioethics, any charters of rights, by stressing the importance of these charters as a *methodological expedient*. In that perspective, the authentic merit of such charters is not related to the specific rights that are recognized therein, but to claims and demands that they justify. These two features are similar at first glance, but deeply different.

In order to explain this statement, we should stress the fact that these statements are of modest value for those who already enjoy the related rights; contrariwise, these charters are highly significant either for those who cannot enjoy them, or for those who consider worthwhile only those rights that are *at the same time* recognized to every human being. This is why declarations of rights and charters are, more than mere lists of rights, list of rights that people can claim. This difference is not entirely negligible; the mere act of claiming means that the one who claims is persuaded that his or her claim deserves to be heard (D'Agostino 1998: 12). What is at stake, here, is not simply a question of subjective perceptions, but of objective status.

If I am persuaded that my claim deserves to be heard, it is because I think that it can be somehow communicated to, and shared with, other people, and that it can be somehow compatible with their respective claims. My subjective intention, my personal purpose, is not enough: in order to lay some demands I have to translate them into objective claims, into something I can communicate and propose within a context of interpersonal dialogue (the public arena): into something that others can understand and eventually endorse. In other words, no one can demand (in the sense of claiming for a public recognition) what is merely useful for himself or herself, nor what is universally important but merely private (e.g. happiness). One can demand only what he or she thinks as being *objectively important and just*.

For this reason, charters of rights are lists of rights and values that people *perceive* as worthwhile and objective: values that are ratified in charters are thus identified neither deductively nor merely through a rational approach, but inductively. That is, starting (from below) from what people concretely perceive – in a specific time – as fundamental for their coexistence and their dignity, rather than starting (from above) from abstract theories or from comprehensive doctrines.

Such an inversion is what we can define as a methodological expedient (D'Agostino 1998:13). It is an expedient, of course, because it is simply a different starting point, a different point of view about what human beings deserve and about what is worthy of honour. But it is useful, for two reasons: first, because it does change the meaning of the agreement, which is perceived as the ground of those rights, and second, because it does include within the public debate even those who can't claim for themselves.

On the one hand, in fact, starting from the concrete perception of what is just and worthwhile makes the eventual agreement on some principles something more significant than a mere procedural outcome. In our pluralistic societies, stressing our respective doctrines, or excessively underlining our anthropological starting points, can be at odds with the possibility of reaching an agreement: for that reason, what we can do is simply to elaborate a fair procedure, and to count votes, thus allocating the final outcome to the majority opinion. Contrariwise, if we stress the *fact* that different people perceive as concretely worthwhile different things, and that they assign a value to different choices, an agreement can take into consideration all these concrete needs instead of the different moral perspectives, in order to verify which can be recognized along with the others, within that particular intersubjective context. In other words, it is a more concrete approach, focused on specific needs and demands instead of on abstract doctrines.

On the other hand, starting from the concrete perception of what is perceived as *objectively* just and worthwhile, such an approach can include in the dialogue even those who can't claim, for their voice is not strong enough, or for their claims are not sophisticated enough (e.g.: rationalized) to participate in the debate. If our discussion is not simply about our ideological perspectives, but it starts from what each of us conceives as objectively just, rights can be granted to every human being, and charters will bind not only those who did participate to the agreement, but even those who didn't. The reason is that we are not recognizing theories but needs, not moral perspectives but concrete demands necessary to protect human dignity: the affirmed rights become the basic needs that are to be recognized to every person, whatever the moral perspective he or she believes (they are somehow analogous to the list of primary goods provided by Rawls, those social conditions that people generally must have in order to pursue their conceptions of the good successfully: Rawls 1993: 75). And this is why these rights can be affirmed even in the name of those who didn't claim anything, or for those who didn't have the possibility to do it (D'Agostino 1998).

Finally, in order to explain why I consider charters of rights important even (or particularly) in the field of bioethics, I can quote E. Lévinas and his statement about the *medical vocation of mankind*: the ongoing movement of rights, in the field of bioethics, is simply our attempt not to leave alone any human being, with his or her loneliness and weakness (Lévinas 1990: 46), in the face of illness or death.

The multicultural challenge

A different question is whether or not such principles and rights can be used to rule bioethical questions, by taking into consideration the diversity of cultures and traditions. In other words, we should discuss whether and how it is possible to define common principles and stable guidelines for different cultural, ethnic, religious and social contexts, and whether or not the Universal Declaration on Bioethics and Human Rights – with particular reference to Article 14 and Article 12 – can be considered a satisfactory and proper answer to such a problem.

A first possible answer is to affirm, on the one hand, that every human being has the right to culture, including the right to enjoy and develop his or her specific cultural life and identity, but that, on the other hand, the right to culture (as any other right) is limited: cultural rights cannot be invoked or interpreted in such a way as to justify the denial or the violation of other human rights and fundamental freedoms, but cultures shall be considered as context in which human rights must be promoted and protected (Ayton-Shenker 1995; Logan 2007).

This approach is fascinating, but it is oversimplifying. First, because rights are always conflicting with each other, and always they are limited by other rights, not only when cultural diversity is at stake. Second, the vagueness of universal principles and rights is precisely what makes these problems possible, because principles can be read, interpreted and enforced in different ways according to different cultural starting points. Third, basic notions to which these principles are referred are themselves highly contested, and susceptible of different understandings: the notions of person, health, autonomy, illness, consent, adulthood, freedom of choice, can be differently understood starting from different points of view. And cultures and traditions are precisely such different points of view about life and death, health and illness, individuality and relationships.

Many works have been devoted to analyze cultural assumptions and traditional perspectives on medicine, health and therapy, and most of them underlined the tension between universal rights and these diverse approaches to health (Thomasma 1997; Bowman 2004; Marshall and Koenig 2004). My aim, here, is to focus on the importance of charters of universal principles, and on legal strategies available to balance – as much as possible – these diverse approaches to medicine and health with the enforcement of rights assumed as universally just and fair.

Diversities are settled in religious, philosophical, traditional, societal grounds, and can produce different interpretations of concepts that are basic elements of universal rights. In that perspective, one can observe an impressively wide list of perspectives, even on these basic notions, which are commonly used and presumed in bioethical debates (Sabatello 2009: chapter 1; Bowman 2004). I can only briefly summarize them, without discussing their roots and their specific characteristics.

- The concepts of illness and medical practice: the Declaration starts from an objective and scientific approach, according to which there are universal grounds and normative standards to describe illness, and to observe and criticize medical practices and their efficacy. On the other hand, for some cultures, the illness and its treatment are attributed to the effect of supernatural forces, or to something that can be neither measured nor observed empirically. For instance, in some Hindu cultures, the bad Karma of a previous life is linked to a present condition of illness, while other cultures adopt a more holistic approach, according to which sickness is the result of mental and bodily factors, as well as of individual and cosmic interactions (Loustaunau and Sobo 1997: 18), and others are focused on the harmony between conflicting principles, such as yin and yang in Chinese medical tradition (Helman 2007: 5).
- Consequently, the role of performer can be legitimately played, in different cultures, by different subjects. While in Western countries those who are responsible for diagnosing and treating can be only professionals with an ascertainable education in medical schools, because their task is scientific and objectively appreciable, in other cultures this role can be played by local healers, family members, or other subjects whose proficiency is recognized from the group. That is not only because their role is, in some cultures, interpreted in a religious or esoteric perspective, but also because their practice is transmitted and learned as an ancestral tradition, something that is not related to formal proficiencies, but to personal expertise and group's recognition (Snow 1974: 93).
- Again, the concept of a sick person can be different: declarations start from an individualistic conception of illness, according to which it always affects a specific, single individual. And this is the reason why the single individual is the one who has to consent to medical treatments, and the one who has to be considered during the whole process of

decision making. But for some cultures personal conditions concern the entire group (the family, the clan, etc.), and thus it is the group, or its spokesperson, who has to take part in the decision-making process. In many Asian cultures, for instance, the patient's family has to be involved in every clinical decision, not because the single person is not practically able to decide about himself or herself, but because illness is something that affects the group, and not only the individual (Cheng-the Tay and Sung 2001: 51).

- Similarly, what makes clinical practices perceivable as beneficial can vary considerably. Practices that are viewed as beneficial and thus highly appreciated within a culture can be contested from a different cultural point of view, and perceived as harmful to others. I'm not referring to highly contested practices, such as euthanasia, which intrinsic meaning and worth are debated within the Western world too, but to practices commonly accepted and appreciated in some perspectives, which are perceived as harmful to others. One interesting example is that of norms of disclosure of information on medical diagnosis: Western countries generally underline the importance of that wide disclosure of information, which is inspired to the principle of truth telling, while other cultures prefer indirect forms of disclosure, to other members of the patient's family group, or forms of disclosure accompanied from religious and non-religious rituals (Candib 2002; Sabatello 2009: 6; Gordon 1995). After all, the extent of such a disclosure is differently appreciated even within Western countries, and in USA higher standards of direct disclosure are accepted than in some European countries.

- Differences are also relevant with regard to the definition of childhood and adulthood, whose dividing line is crucial in medical practice when personal autonomy and self-determination about treatments are at stake. Although children are generally assumed as lacking the rationality and the freedom to express their will, the extent of such a group is different among cultures. Consequently, the moment when adults can (or shall) express their will, and choose what they consider the best interest for the child, should change considerably.

- Related to this crossroads is the understanding of the human body: that is, the question of its meaning and its worth, and the question of the legitimacy of ritual bodily practices, which are grounded in long-standing traditions and beliefs. Western societies, and contemporary charters of rights, start from an individualistic point of view, according to which bodies are personal belongings, partially or fully available for the one who owns them. Other cultures stress the fact that the body is the place where personal identities are rooted, but also where the group's identity, personal belonging and individual integration are connected with each other. For this reason, bodily practices have a meaning that is differently perceived, and that is connected to the social value associated to the human body (Durkheim 1965: 472 ff.).

- In that perspective, health and self-care practices can vary considerably. Behaviours related to personal hygiene, diet, exercise, illness-avoidance practices are highly influenced from culture and the cultural environment. Dietary self-care practices are highly influenced by culture, and determine how much and what kind of food shall be eaten, when special diets are needed, balance among different kinds of food and foods to avoid. Similarly, culture influences what kind of hygienic practices are recommended or compelled, and what special cares are due to specific body parts (eyes, teeth, genitals, etc.) (Bonder 2013: 112).

- Finally (but that list is far from being exhaustive) the legal relevance of the practices involving health, food, hygiene, self-care leads us to take into account the question of the rights holder. The Declaration, such as the majority of similar documents, considers the single individual as the main figure of rights' holder, the one who has to be protected and granted. But when we consider the worth that is assigned to dietary or self-care practices by different cultures, the role of groups becomes pivotal. In other words, the question is whether or not, and to what extent, groups can demand *as such* to be endowed with the protection and exercise of rights; the problem is, as I will show, that on the one hand charters and declarations do protect "respect for cultural diversity and pluralism" (Declaration Article 12), but on the other hand they adopt an individualistic framework of rights and guarantees. And it is evident that there is the risk of a clash, for instance, between the collective worth of traditional health or bodily practices, and individual rights, if both are sanctioned and protected.

Bioethical principles and cultural discretion

In order to disentangle such a muddle, we need to start from the assumption that what is at stake is not the philosophical alternative between universalism and cultural relativism. I consider cultural relativism, according to a long-time philosophical tradition, self-contradicting and inadequate. It is self-contradicting because it is true that different moral principles exist, the relativism states, at the same time, the universality of the relativist principle, by stating the absolute prescription that all prescriptions are relative (Teson 1985; Dundes Renteln 1990). Additionally, it is inadequate, because it cannot demonstrate the necessity of tolerance or other ways to accommodation among different perspectives: in other words, it does not clarify why we should tolerate or respect practices prescribed by unfamiliar moral/cultural systems, because the mere observation of cultural diversity can't be understood as a prescription of what ought to be (Hatch 1983: 67). I admit that many other arguments should be proposed in order to criticize cultural relativism (Tilley 2000: 501), but for the moment it can be enough to say that this perspective is at least questionable.

On the other hand, we should stress the fact that any declaration of rights, and the Declaration on Bioethics and Human Rights is no exception, is a compromise, the fruit of several ideologies and of diverse conceptions of human beings and society. They constitute, to some extent, an "adjustment" to a multicultural and divided world of local conceptions of human dignity. To some extent, they try to fulfil the hope expressed in 1947 by the Chinese delegate to the UN Commission on Human Rights, to reconcile Confucius and Thomas Aquinas. (Macmillan 2003). As rightly pointed out (ten Have 2013; Kirby 2009a; Revel 2009), the scope of the Declaration is in itself, at least in part, that of being a compromise, in order to addresses "ethical issues related to medicine, life sciences and associated technologies as applied to human beings, taking into account their social, legal and environmental dimensions" (Article 1a): to take into account cultures and environments, thus, is not a choice, nor a mere opportunity, but a pivotal dimension of any international charter of rights, such as the Declaration on Bioethics and Human Rights.

What is at stake, here, is the possibility of a reasonable accommodation of cultural traditions, taking into account the different impact that some rights have in particular contexts because of cultural or religious or ethnic considerations. That reasonable accommodation means simply that these factors (which we can call the cultural framework) shall in some cases be taken into consideration when deciding whether or not there is a breach of the international human rights (and bioethical) standard: any specific standard, or many of them, can impact differently in different local situations (de Varennes 2006: 79).

Each culture – according to the UNESCO Universal Declaration on Cultural Diversity – "should be regarded as a set of distinctive spiritual, material, intellectual and emotional

features of society or a social group and that it encompasses . . . lifestyles, ways of living together, value systems, traditions and beliefs". On the one hand, cultures shall be
respected and taken into account, but on the other hand such considerations shall not be
invoked to infringe human rights and fundamental freedoms nor to limit the scope of the
principles set out in the Declaration. It is interesting to notice that in a preliminary draft
elaborated by IBC, Article 7 stated that "Any decision or practice shall take into account the
cultural backgrounds, school of thoughts, value systems, traditions, religious and spiritual
beliefs and other relevant features of society", while in its final draft (Article 12) any reference to decision or practice, as well as the mentioning of backgrounds, schools of thought,
traditions, and beliefs, has been removed (Revel 2009).

Accordingly, the principles of the Declaration shall be considered jointly, as interrelated
and complementary (Article 26). The principle of cultural recognition cannot simply be
let aside, but it shall be considered as representing the context of other principles. This
perspective, I argue, can reduce the ambivalence that stems from the dual recognition of
cultural diversity and universal human rights, showing how principles can be conflicting, of
course, but how they can also be harmonized within a common framework.

Such a kind of harmony, evidently, is not easy to reach. What I suggest is to adopt a strategy of interpretation, such as to realize at the hermeneutical level some compromises with
universalistic claims. In other words, what I suggest is to realize compromises during the
implementation of the principles expressed by the Declaration, similar to those which have
been achieved during its drafting.

But what does it mean, concretely? How can this approach work, and how can it deal
with bioethical issues?

I argue that it is possible to adopt a multifaceted approach, in order to balance the
enforcement of universal standards and principles, with the recognition of local traditions
and cultures. In that perspective, many different strategies can be implemented, but none
of them can be established as the best one in each situation: on the contrary, each of them
can be useful to balance universal principles and cultural traditions, depending on specific
characteristics of the case and on subjects involved.

For instance, one first question is to establish whether such a balance is settled on a
sub-national or international level. If the problem is the clash between universal principles,
which have been recognized by a national legal system, and norms that arise from a minority group's tradition, it would be possible to adopt a strategy somewhat similar to the doctrine of reasonable accommodation that Canadian courts have adopted since the mid-1980s
(see *Ontario Human Rights Commission v Simpson Sears* 1985), but providing a stronger
protection of human rights through a more specific hermeneutical procedure. The Canadian strategy (see Gaudreault-DesBiens 2009: 152)

> allows an individual who is detrimentally affected by an otherwise neutral norm the
> possibility to require, as a matter of law, to be accommodated. This accommodation,
> which essentially consists in the bending of an existing norm or in the creation of a
> particularized regime for the claimant (whether through an exemption or through a
> specific permission to do something), can only be refused if it imposes an undue hard-
> ship upon the organization from which the accommodation is requested.

The strategy I'm arguing for can be defined as "contingent accommodation": through
that strategy, the group's authority and traditional rules can be recognized about some specific life events (such as rites of passage) crucial to the group's identity, or in certain social

arenas (dietary or hygienic practices, funerary rituals, animal slaughtering), or in particular ways of implementing universal principles. According to this last option, for instance, the patient's culture should be taken into account when his or her informed consent is due, carrying out a strategy suited to this particular point of view. In other words, because personal culture can influence both the quantity of information, the way of giving them, and the persons who have to be involved, all these factors can deeply change the meaning of informed consent and of practices which are consistent to it (Bowen 2008).

Outside these specific situations, the authority of state law shall be affirmed. In other words, apart these exceptions, universal principles shall be enforced by state law, and group autonomy is allowed only as long as they respect some minimal state-defined standards. If they don't, the state can override them by applying its rules, so protecting the group's vulnerable members when their interests are violated: for example, groups are allowed to establish rituals in adulthood passage, provided they do not violate some minimal standards regarding bodily integrity and personal freedom. Moreover, groups (their spokesperson, whether there are, or directly single members for specific situations) could ask for a recognition of traditional health care practices, provided they fulfil some minimal standards: for instance, parents might be free to take care of their children by following traditional medicine or by entrusting to a performer whose proficiency is recognized within the group, provided they do follow some minimal standards established by law. These standards could be about medicines which are allowed and which are not, about dietary, about physical growth, about vaccines, about hygiene, but as I said, they should be minimal, thereby allowing a space for such a cultural accommodation.

That approach is similar to Shachar's transformative accommodation (Shachar 2001: 127). From that perspective what is pivotal is the idea of a system of ongoing mutual adjustments by both the group and the state (the larger community), and the idea that every solution shall be continuously reshaped and adapted during the time, taking seriously into account the internal development of cultures and traditions. At the same time, this ongoing dialogue can encourage traditional groups to re-examine the elements of their tradition, because the group's members put pressure on the group's authorities for changing internal rules, if these rules appear to be unjustly discriminating or overwhelming compared to the general law's rules.

However, if what is at stake is the culture and traditional rules of a larger community, such an approach doesn't work. In other words, a different strategy may be adopted when the challenge is the adoption of universal principles (such as those sanctioned by the Declaration) within a community that starts from different conceptions of the basic elements of bioethical principles.

A margin of appreciation for bioethical principles

A promising approach is a system of legal interpretation, somewhat similar to the doctrine of the margin of appreciation, which the European Court of Human Rights adopts in its case law to balance rights' enforcement and national traditions (see, among others, Arai-Takahashi 2001; MacDonald 1993; for a recent and comprehensive work, see Legg 2012).

I argue that allowing restrictive interpretations of the Declaration's principles should be possible, so as to limit their goals and validity, in order to balance them with the right to cultural identity. Of course, this "culturally sensitive" interpretation cannot be too wide: in particular, universal principles can neither be limited without justification, nor can their interpretation be such as to abolish them or to determine an application incompatible with their nature. Besides this, two more caveats shall be considered in order to determine the width of such interpretation: first, it must be taken into account the specific nature of rights involved, and second, it shall be considered the existing consensus on the specific issue or its problematic status. Let me explain.

If we adapt this doctrine to bioethical issues, I argue that it is possible to say that while some rights cannot be limited, other rights may be, both by defining them and by restricting their exercise, for reasons such as the protection of national security and of public morals, reasons that are always related to the place, and that each single state can better appreciate. Of course, this appreciation can never be as strong as to unreasonably compress a right, or to abolish it, or to determine an application incompatible with its very nature. However, and at least with regard to certain rights, it can lead to very different applications from one state to another.

But what kind of rights can be limited, and what cannot? According to some scholars, it is possible to scale absolute rights (life, prohibition of torture), strong or qualified rights (fair trial; liberty; derogations; privacy; freedoms of religion, assembly and speech; and non-discrimination) and "weak" rights (property, education and free elections), progressively affecting the width of the margin of appreciation and the amount of deference to be accorded to the state (Legg 2012: 200). If we adapt that distinction to bioethics, we can distinguish strong rights (such as equality, justice and equity; access to quality health care and essential medicines; access to adequate nutrition and water; improvement of living conditions and the environment; elimination of the marginalization and the exclusion of persons on the basis of any grounds; the priority of an individual's interests and welfare over the sole interest of science or society; etc.), qualified rights (prior, free and informed consent of the person concerned to medical interventions or research; respect of human vulnerability; non-discrimination and non-stigmatization; respect for cultural diversity and pluralism; protection of future generations and environment; etc.), and week rights (solidarity among human beings and international cooperation; promotion of health and social development for their people as a central purpose of governments; the sharing of benefits resulting from any scientific research and its applications; respect of bodily integrity; etc.).

Of course, that list is highly questionable, in the sense that one can challenge my choice of considering one specific right as strong or week. But what is at stake, here, is not the status of every single right, but the idea that, among different rights, some are stronger than others. This distinction simply means that for some rights (which I called as strong rights) no argument is solid enough to justify any kind of limitation, while other rights (qualified and weak) may be, whether by defining them or by restricting their exercise, for reasons such as the protection of a cultural identity or as the diverse meaning that is given to a specific concept, which is implicit in the right.

A second parameter is that of consensus: the width of such a cultural interpretation should be defined considering the existence of a consensus between different state rules and enforcement traditions. More specifically, the width of interpretation is inversely proportional to the consensus on a specific perspective or on the meaning of a specific concept (Benvenisti 1999): the more we are able to identify, on the international arena, a wide consensus on a particular issue, the less this margin of interpretation can be granted to national institutions claiming for traditional practices.

In other words, it is possible to say that in bioethical issues a margin of interpretation should be granted as much as one can observe a remarkable diversity of moral and legal rules about a specific issue. For instance, there is undoubtedly a wider consensus about the ban of human cloning than about euthanasia, and great disagreements can be observed about the meaning and the width of personal self-determination; again, there is a wider consensus on the principle of access to health care, medicines, nutrition and water than about limits of animal testing or about dietary prescriptions.

The main problem of this approach is that the margin of appreciation doctrine works, in ECHR case law, as a judicial argument, and only with regard to specific cases and concrete situations; however, this is not the case with regard to the Universal Declaration of Bioethics and Human Rights. The Declaration is a non-binding document: like other declarations adopted by UN agencies, this document is what we can call a soft law instrument (this topic will be further developed in Part II of this volume), something that is weaker than conventions because it is not intended to oblige states to enact enforceable rules inspired by common standards, but to encourage them to do so (Andorno 2007). Its adoption by the General Conference of UNESCO does not make it part of the international law, and it is not a treaty open to ratification by states or international organizations, nor does it bind Member States to conform to its provisions. The Declaration is "hortatory, aspirational and educational rather than legally normative" (Kirby 2009a: 73). For this reason, there is no international Court directly devoted to its interpretation and enforcement.

It does not mean, however, that the proposed mechanism of interpretation cannot work at all. It means that the same mechanism can work differently as the ECHR does, with regard to the different nature of the Declaration. In fact, it should be considered that policies and principles expressed in a non-binding document by a United Nation agency can influence the international discourse and stimulate the evolution of the customary international law. This is what happened to the UDHR of 1948, that has undoubtedly informed the reasoning of judges in the intervening years, both in international and national courts and bodies. This

> does not mean that, as such, the principles of the UDHR bind the judge or state the content of a legal rule in the way a municipal rule would. It simply means that, by offering a general principle, accepted by an organ of the international community, the judge is afforded a mooring, or bearings, for an approach to the case in hand.
>
> (Kirby 2009b: 329)

Thus, the problem is: how should we interpret the Declaration, and what space can we recognize – as interpreters – to the cultural context?

On the one hand, the Declaration principles should be interpreted in the same way other treaties are interpreted in international law, by having regard to Article 31 of the Vienna Convention on the Law of Treaties. According to this rule, such treaties are to be interpreted (Article 31.1) "in good faith in accordance with the ordinary meaning to be given to the terms of the Treaty in their context and in the light of its object and purpose"; the reference to the context permits to take into account both subsequent agreement and practices regarding interpretation, and even to consider the recent shift towards purposive approaches, which has been adopted in many national jurisdictions (Kirby 2009a: 74).

On the other hand, the need for a liberal and purposive approach to interpretation is stronger with regard to documents drafted in multiple languages, through the involvement of experts of multiple legal traditions and cultures, and designated to promote desirable objectives and goals. In other words, if the starting point of any interpretation remains a textual analysis, the role played by the states in order to give effect to the principles set out in the Declaration should be stressed (Article 22). I suggest that such a role might be intended not only as a call to the Legislator and to the Executive to promulgate rules and any other "appropriate measure" (Article 22, 1), but also as a reference to ethics committees and to the Judiciary, in order to play a role in interpreting the Declaration and in the assessment of its rules and principles (Boussard 2009).

If we, once more, are mindful of the fact that the main worth of the Declaration is not merely nor mainly legal, but epistemological, we can argue that a wide margin of interpretation is perfectly sound with the Declaration's purposes and aims. In other words, if we take seriously the aims of (Article 2) *guiding* "States in the formulation of their legislation, policies or other instruments", of *promoting* "respect for human dignity", of *fostering* a "pluralistic dialogue about bioethical issues", a purposive hermeneutic and a wide margin of interpretation of the Declaration is highly appropriate, not least because of its non-mandatory language. In such a wide margin of interpretation, I argue that is possible and wise to take into account the culture and traditions of the social context where principles and rules are to be implemented.

In this perspective, if we want to recognize a margin of interpretation to single cultural traditions with regard to the bioethical principles, we should recognize that some practices, which from our point of view seem to be at odds with some principles, can be consistent with these principles from different cultural starting points. It is, in other words, a problem of recognition, more than directly or essentially a legal question.

We should recognize that those principles, which we usually interpret in a certain way, can be differently understood by other cultures, and that they *can do so legitimately*: because they can claim a margin of interpretation as a legitimate hermeneutical parameter, and such a criterion is what makes it possible to balance universal principles with local cultural traditions, at least within the international debate.

For instance, the principle of personal autonomy may be interpreted, from different cultures, in different ways, which take into account traditional conceptions of childhood and adulthood. The problem is that depending on the beginning of the adult age, a person is allowed to legitimately dispose of his or her body, to be considered in health care decisions, and to express a valid consensus. But the meaning of adulthood and childhood is clearly a cultural question, at least in part, thus it would be possible to allow different interpretations of these words, and to recognize them as consistent with universal standards of the Declaration: in other words, it may be possible to recognize different practices as consistent with the same international principle, provided that such interpretation does not imply a complete denial of this concept: e.g., the adult age can never begin before the physical development is complete.

Concretely, this mechanism can work as an internationally accepted argument, according to which different communities, starting from different cultural points of view, can accept and endorse international charters of rights (e.g. the Declaration), without worries about the giving up of their culture. Those documents, although closely related to the Western tradition, do not necessarily express a cultural imperialism, nor are they necessarily inconsistent with local traditions. Any culture can legitimately interpret the Declaration in a way that the stated principles become as much as possible consistent with local culture, so to be perceived as *internal* to it (or, at least, not too far from it). In this perspective, no international body nor any international authority can criticize such different interpretations of universal principles, provided these interpretations do not unreasonably compress a right, nor do they determine an application incompatible with its very nature. Being an hermeneutical parameter, this cultural sensitivity does not affirm that any interpretation is sound, but simply that different applications from one culture to another are possible, at least with regard to certain principles (Macioce 2014, 2016a, 2016b).

The basic idea of such a perspective is purely hermeneutical: truth, and human dignity, are regulative principles that play the role of guiding our search (and our judicial activity) under the assumption that we can make mistakes and learn from them, in an ongoing process of dialogue and adjustments. Similarly, according to Finnis (Finnis 1980: 231–233), we can therefore distinguish between concepts of rights (e.g. of the human right to life, of a fair trial), which can be widely shared, and conceptions of rights, which can deeply differ from each other. In a similar perspective, Michael Walzer (Walzer 1994) distinguished between minimal and maximal meanings of moral concepts, between 'thin' and 'thick' accounts of morality; both are useful at different times and for different purposes, and they can interact and work in conjunction. In other words, this is a way to concretely perceive the difference between universality (of rights) and uniformity (of rules) (Davidson 2001).

Let me give an example. If we consider Articles 5, 6 and 7 of the Declaration we find a notion of informed consent based on a specific conception of personal autonomy. According to this conception, the individual is seen as the one who can make medical decisions on his or her own behalf. This is because the Western culture appreciates individual liberty and self-determination as the more significant values that should be taken into account, and because illness is interpreted both as something that affects exclusively the individual body (except for specific cases), and as something that concerns the individual experience. Thus, this model requires medical providers both to give patients all the information they need to decide about treatments, and to respect what patients intelligently and voluntarily decided: in other words, this principle does compel a disclosure of information to patients as wide as possible, and the respect for patient's self-determination about treatments to be performed or avoided.

However, people from several cultures do not really ask for a complete and autonomous control over their medical decision making. At the same time, rules governing conversations between physicians and patients do not take into account that according to their culture some patients want to involve different decision makers, and they ignore the fact that not all patients want the same information content in disclosures, and not all information is sought for the purpose of medical decision making (Bowen 2008; Blackhall et al. 1995). For these reasons, the principle of autonomy and individual responsibility, such as that of informed consent, can be interpreted as allowing different contents in disclosures, and different procedures of decision making: for instance, in order to acknowledge the real influence of cultural context in which the patient is embedded, family-centred decision-making procedures should be allowed, as well as forms of family group involvement (Marshall 2000; Shaibu 2007; Bonder 2013: 93 ff.). In addressing these peculiar perspectives, we can stress the fact that from a same concept of autonomy (like that affirmed by the Declaration) we can infer two different conceptions of personal autonomy, and one of them can be much more congruent with the patient's culture: for instance, through a more communitarian understanding of individual desires, plans, values, we can conceive them as not independent from the relational network within which the person is embedded, but related to them.

In a similar way we can address the question of bodily integrity and of traditional practices involving children's or adolescents' bodies, which Article 8 of the Declaration (Respect for human vulnerability and personal integrity) protects by stating that "Individuals and groups of special vulnerability should be protected and the personal integrity of such individuals respected". For it is unquestionable that minors and children are subjects of special vulnerability, we can consider cases in which they are subjected to practices such as ritual scarification or other mutilations.

In the much-publicized English case *R. v Adesanya* (1974), for instance, a woman from Nigeria, claiming she was following her traditional customs, made small incisions with a razor blade on the faces of her two sons, aged nine and fourteen: the children were said to be willing, and their mother's intention was to make their sons perceived as belonging to the Yoruba culture. Even if the judge granted the mother an absolute discharge, understanding that she did not realize she was breaking the law, I argue that this case, and all similar ones, should be ruled not by making exceptions in the name of personal ignorance of the law, but by the recognition of a possible, different interpretation of the bodily integrity. In other words, it is not a matter of ignorance, but of diverse conceptions of the same concept.

In that case, the correctness of a different interpretation of bodily integrity should be recognized, even if such an interpretation is strongly far from the mainstream (Western) bioethical culture. It is possible, for instance, that according to some cultures physical integrity can be respected not merely when the human body is completely and perfectly intact, but also when a small and not dangerous mutilation contributes to its "purity", according to the hermeneutic horizon within which the ritual practice is acted. Of course, this interpretation cannot be as wide as to imply a complete denial of the concept, e.g. by allowing mutilations that can represent a serious danger for the individual's health – this is evidently the case of female genital mutilations.

Conclusion

Global bioethics is a two-level phenomenon. The setting of fundamental values on which traditions and cultures agree is to be balanced by the efforts to articulate more concrete bioethical standards within the context of specific religious and cultural traditions. And according to some scholars, it is precisely the dialectic of global and local perspectives that can help to construct and corroborate a global bioethics (ten Have and Bert 2013).

Therefore, an anthropological approach to health care and bioethics, and a more nuanced framework of interpreting bioethical principles, can avoid many weaknesses of perspectives that presume the existence of rigid and commonly understandable values. As I tried to explain, this more culturally sensitive perspective does not disavow all claims to common minimal standards in bioethics; rather, it simply call us to take seriously "the multiplicity of modes of practical moral reasoning", and push the bioethics to take an important step in acknowledging the "multiplicity of moral worlds" (Turner 2001) without falling into the shortcomings of moral relativism.

In this perspective, I've argued that a global community of shared principles should be fostered, as the outcome of an ongoing process of interpretation, negotiation and dialogue. These principles cannot override the diversity of cultures and traditions, but – at least in part – they can make these traditions compatible with each other through a process of interpretation. Of course, interpretation will never eliminate ethical disagreements, but it will reduce the distance between different starting points that derive from different cultural traditions. In other words, the setting of common principles at the international level can proceed along with the recognition of deep cultural diversities.

As Parekh pointed out, cultural diversity is not (or not primarily) about conflicting values, but about universal values (even those sanctioned in UNESCO's Declaration) that need to be interpreted, prioritized and eventually reconciled, in the light of specific circumstances of each society (Parekh 2006: 128).

References

Alexy, Robert 2002, *A Theory of Constitutional Rights*, trans. J. Rivers, Oxford: Oxford University Press.

Andorno, Roberto 2007, Global Bioethics at UNESCO: In Defence of the Universal Declaration on Bioethics and Human Rights, *Journal of Medical Ethics*, 33: 150–154.

Annas, George J. 2004, American Bioethics and Human Rights: The End of All Our Exploring, *The Journal of Law, Medicine & Ethics*, 32: 658–663.

Arai-Takahashi, Yutaka 2001, The Defensibility of the Margin of Appreciation Doctrine in the ECHR: Value-Pluralism in European Integration, *Révue Européenne de Droit Public*, 3: 1162.

Ayton-Shenker, Diana 1995, *The Challenge of Human Rights and Cultural Diversity, United Nations Background Note*, United Nations Department of Public Information DPI/1627/HR-March 1995 <www.gem-ngo.org/uploads/Subtheme_A.doc>.

Benatar, David 2005, The Trouble with Universal Declarations, *Developing World Bioethics*, 5: 220–224.

Benvenisti, Eyal 1999, Margin of Appreciation, Consensus and Universal Standards, *International Law and Politics*, 31: 850.

Berlinguer, Giovanni, De Castro, Leonardo 2003, *Report of the IBC on the Possibility of Elaborating a Universal Instrument on Bioethics*, UNESCO 13.06.2003, SHS-2004/DECLAR.BIO-ETHIQUE CIB/6.

Blackhall, Leslie, Murphy, Sheila, Frank, Gelya, Michel, Vicki, Azen, Stanley 1995, Ethnicity and Attitudes Toward Patient Autonomy, *Journal of American Medical Association*, 274(10): 820–825.

Bonder, Bette 2013, *Culture in Clinical Care: Strategies for Competence*, 2nd Edition, Thorofare, NJ: SLACK Incorporated.

Boussard, Hernandez 2009, Article 22: The Role of States, in *The UNESCO Universal Declaration of Human Rights: Backgrounds, Principles and Application* (pp. 293–302), Paris: UNESCO Publishing.

Bowen, Matthew D. 2008, Race, Religion and Informed Consent: Lessons from Social Science, *Journal of Law, Medicine & Ethics*, 36: 150.

Bowman, Kerry 2004, What Are the Limits of Bioethics in a Culturally Pluralistic Society?, *Journal of Law, Medicine & Ethics*, 32: 664–669.

Candib, Lucy 2002, Truth Telling and Advance Planning at the End of Life: Problems with Autonomy in a Multicultural World, *Family Systems & Health*, 20: 213–228.

Cheng-the Tay, Mark, Sung, Lin 2001, Developing a Culturally Bioethics for Asian People, *Journal of Medical Ethics*, 27(1): 51–54.

D'Agostino, Francesco 1998, *Bioetica, nella prospettiva della filosofia del diritto*, Torino: Giappichelli.

Davidson, Scott 2001, Human Rights, Universality and Cultural Relativity: In Search of a Middle Way, *Human Rights Law and Practice*, 6: 97.

de Varennes, Fernand 2006, The Fallacies in the "Universalism versus Cultural Relativism" Debate in Human Rights Law, *Asia Pacific Journal on Human Rights and the Law*, 7(1): 67–84.

Dundes Renteln, Alison 1990, *International Human Rights: Universalism versus Relativism*, New York: Sage Publications.

Durkheim, Emile 1965, *Elementary Forms of Religious Life*, New York: Free Press.

Faunce, Thomas 2014, Bioethics and Human Rights, in *Handbook of Global Bioethics* (pp. 467–484) ten Have, Henk, Gordijn, Bert, eds., Netherlands: Springer.

Finnis, John 1980, *Natural Law and Natural Rights*, Oxford: Clarendon Press.

Gaudreault-DesBiens, Jean-François 2009, Religious Challenges to the Secularized Identity of an Insecure Polity: A Tentative Sociology of Québec's 'Reasonable Accommodation' Debate, in *Legal Practice and Cultural Diversity* (pp. 151–274) Grillo, Ralph, ed., Farnham: Ashgate.

Gordon, Edwin 1995, Multiculturalism in Medical Decision-Making: The Notion of Informed Waiver, *Fordham Urban Law Journal*, 23(4): 1321–1362.

Gross, Hyman 1969, Standards as Law, *Annual Survey of American Law*, 34: 66–78.

Hatch, Elvin 1983, *Culture and Morality*, New York: Columbia University Press.

Helman, Cecil H. 2007, *Culture, Health and Illness*, London: Hodder Arnold Press.

Kirby, Mike 2009a, Article 1: Scope, in *The UNESCO Universal Declaration of Human Rights: Backgrounds, Principles and Application* (pp. 67–81), Paris: UNESCO Publishing.

Kirby, Mike 2009b, Human Rights and Bioethics: The Universal Declaration of Human Rights and UNESCO Universal Declaration of Bioethics and Human Rights, *Journal of Contemporary Health Law & Policy*, 25: 309–331.

Langlois, Adele 2013, *Negotiating Bioethics: The Governance of UNESCO's Bioethics Programme*, London: Routledge.

Legg, Andrew 2012, *The Margin of Appreciation in International Human Rights Law: Deference and Proportionality*, Oxford: Oxford University Press.

Lévinas, Emmanuel 1990, Ethique et transcendance. Entretiens avec le philosophe Emmanuel Lévinas, in *Médicine et éthique: Le dévoir d'humanité* (p. 475) Hirsch, Emmanuel (ed.), Paris: Le Cerf.

Logan, William S. 2007, Closing Pandora's Box: Human Rights Conundrums in Cultural Heritage Protection, in *Cultural Heritage and Human Rights* (pp. 33–52) Silverman, Helaine, Ruggles, D. Fairchild, eds., New York: Springer.

Loustaunau, Martha O., Sobo, Elisa J. 1997, *The Cultural Context of Health, Illness, and Medicine*, Westport, CT: Bergin & Garvey.

Macioce, Fabio 2014, *Il Nuovo Noi. Immigrazione e integrazione come problemi di giustizia*, Torino: Giappichelli.

Macioce, Fabio 2016a, Balancing Cultural Pluralism and Universal Bioethical Standards: A Multiple Strategy, *Medicine, Health Care and Philosophy*, 19: 393–402.

Macioce, Fabio 2016b, A Multifaceted Approach to Legal Pluralism and Ethno-Cultural Diversity, *The Journal of Legal Pluralism and Unofficial Law*, 48(1): 1–16.

Macklin, Ruth 2003, Dignity Is a Useless Concept, *British Medical Journal*, 327, 1419–1420.

Macklin, Ruth 2005, Yet another Guideline? The UNESCO Draft Declaration, *Developing World Bioethics*, 5: 244–250.

Macmillan, Margaret 2003, *Paris 1919: Six Months That Changed the World*, New York: Random House.

Marshall, Patricia 2000, Informed Consent in International Health Research: Cultural Influences on Communication, in *Biomedical Research Ethics: Updating International Guidelines: A Consultation* (pp. 100–134) Levine, R. J., Gorovitz, S., Gallagher, J., eds., Geneva, Switzerland 15–17 March, 2000. Geneva: Council for International Organizations of Medical Sciences.

Marshall, Patricia, Koenig, Barbara 2004, Accounting for Culture in a Globalized Bioethics, *Journal of Law, Medicine and Ethics*, 32: 252–266.

Parekh, Bhikhu 2006, *Rethinking Multiculturalism: Cultural Diversity and Political Theory*, London: Macmillan.

Rawls, John 1993, *Political Liberalism*, New York: Columbia University Press.

Sabatello, Maya 2009, *Children's Bioethics: The International Biopolitical Discourse on Harmful Traditional Practices and the Right of the Child to Cultural Identity*, Leiden, NLD: Martinus Nijhoff.

Shachar, Ayelet 2001, *Multicultural Jurisdictions: Cultural Differences and Women's Rights*, Port Chester, NY: Cambridge University Press.

Shaibu, Sheila 2007, Ethical and Cultural Considerations in Informed Consent in Botswana, *Nursing Ethics*, 14(4): 503–509.

Smith, George P. 2005, Human Rights and Bioethics: Formulating a Universal Right to Health, Health Care, or Health Protection?, *Vanderblit Journal of Transnational Law*, 38: 1295.

Sunstein, Cass R. 1995, Incompletely Theorized Agreements, *Harvard Law Review*, 108(7) (May, 1995): 1733–1772.

Sweet, William, Masciulli, Joseph 2011, Biotechnologies and Human Dignity, Bulletin of Science, *Technology & Society*, 31(1): 6–16.

ten Have, Henk 2013, Global Bioethics: Transnational Experiences and Islamic Bioethics, *Zygon*, 48(3): 600–617.

ten Have, Henk, Gordijn, Bert 2013, Global Bioethics, in *Compendium and Atlas of Global Bioethics* ten Have, Henk, Gordijn, Bert, eds., Berlin: Springer.

ten Have, Henk, Jean, Michèle S. (eds.) 2009, Introduction, in *The UNESCO Universal Declaration of Human Rights: Backgrounds, Principles and Application* (pp. 18–56), Paris: UNESCO Publishing.

Teson, Fernando 1985, International Human Rights and Cultural Relativism, *Virginia Journal of International Law*, 25: 869–886.

Thomasma, David C. 1997, Bioethics and International Human Rights, *Journal of Law, Medicine & Ethics*, 25: 295–306.

Thomasma, David C. 2008, Evolving Bioethics and International Human Rights, in *Autonomy and Human Rights in Health Care: An International Perspective* (pp. 11–24) Weisstub, David N., Díaz Pintos, Guillermo, eds., Netherlands: Springer.

Turner, Leigh 2001, Medical Ethics in a Multicultural Society, *Journal of the Royal Society of Medicine*, 94: 592.

Walzer, Michael 1994, *Thick and Thin: Moral Argument at Home and Abroad*, Notre Dame, IN: University of Notre Dame Press.

V Instances of values and historicity

The philosophical thread of the Universal Declaration on Bioethics and Human Rights

Emilia D'Antuono and Emilia Taglialatela

1. Introduction

There are several points of view starting from which it is possible to analyze the concepts related to the statements' thread of the Universal Declaration on Bioethics and Human Rights (UNESCO 2005). Such concepts can be considered as the cornerstones of ethical, political, social and juridical major issues that have increasingly animated the international debate questioning the practical reason on the impervious ground of the search of more and more democratic ways to regulate the progress of technosciences.

The development of bio-medicine and, more recently, of the so-called convergent technology, to be considered within the wide frame of their social and institutional implications, demands commitment to shared principles, procedures of dialogue and decision-making, individual and collective responsibilities regarding the orientation of transformative processes aimed at the promotion of justice and equity in the present and in the future.

Considering the long and winding path of the attempts to overcome ideological contrasts and create opportunities for an inclusive debate and an open deliberation to acknowledge differences and pluralities, the UNESCO Declaration of 2005 represents a milestone for its intent to propose a gripping connection between the problems emerging from the bioethical debate and the extended constellation of themes resulting from the reflection and the practice of human rights. Such connection is not new, however, since the genesis of bioethics is based on the urgency of critical thought about the modalities of intervention on living things taking into account the epochal changes determined by the "dilatation of possible" (D'Antuono 2014: 197) unceasingly renewed by science and technology. Bioethics does not start from scratch or grows in the void; indeed, according to its original context and its development, this field of research requires the comprehension of the intrinsic cross-reference to the formulation of human rights.

The early beginning of bioethics was marked by the need to regulate medical trials, which led to a special attention paid to human rights in order to avoid discrimination and violence. Such attention remains a feature of bioethical inquiries and practices, finding evidence in further phases of their developments: starting from the evolution of bioethical investigations in the 1960s and 1970s, with the discovery of new therapeutic techniques like transplants, and thus the redefinition of the criteria for death, or the more recent progress of genetic engineering, molecular biology and synthetic biology, which elicit the present debates on hypothetical scenarios of the post- and trans-human. As time went by, bioethics established itself as a philosophical, juridical, ethical and political subject of study involving all the opportunities and risks originating from more advanced forms of intervention on life, with a special focus on the autonomy, dignity and integrity of the human being, including personal freedom and rights in relation to any external power.

The preamble to the UNESCO Declaration shows that its primary aim was to root bioethics within the legal frame of the international law of human rights. The reference to regional and international instruments specifically dedicated to bioethical questions is integrated by a wide range of documents through which the United Nations, since its Universal Declaration of 1948, has promoted the emancipation from any form of discrimination, exclusion or exploitation. The whole body of issues mentioned in the Declaration witnesses the acquisition of the full awareness of the implications connected to the progress of science and technology. Such awareness also involves the role of communication and education as the foundation of citizenship and active participation in the public debate, as well as a different view on the concept of space and time necessarily connected to the currently globalized scientific research and the relation between present and future. The present time must lead to a future that will still endure human life, which the UNESCO Declaration considered not as an unchangeable natural fact but rather as a historically dynamic one, demanding an effort for the comprehension of the transformations continuously caused by science and technology.

In this respect, the Declaration contributes to increase the full awareness that the future is linked to humankind's knowledge and action and to the evaluation of the practice resulting from them.

2. The re-semantization of universality: the contribution of the UNESCO Declaration

Bioethics outlines itself as a field of research that can highlight new connections among scientific, philosophical, ethical, juridical, economical and psycho-social disciplines; within these connections it creates a fresh point of view on themes and problems regarding life and its persistence. Bioethics, then, "looks at" people and living worlds focusing on the individuals' existence and their multiple types of relations. Bioethical themes involve every person and the whole humankind in relation to the bios and the cosmos, which are supposed to inquire, to find solutions and take on responsibilities in unison.

One of the most important elements of innovation of bioethics is the establishment of public debates concerning interpretative categories meant to increase a fully democratic ethos. As a result, such debates specify new meanings of citizenship through the definition of new rights and new forms of participation on the themes of life.

Plurality, then, seems to be bioethics' main feature: plurality of knowledge, through contaminations of competences exploring the mobile frontiers of technoscientific research; plurality of social, institutional and political figures who are supposed to question themselves on such mobile frontiers and make decisions regarding the modalities of governance of the transformation process underway.

It is important to underline that this process not only involves exceptional situations related to specific possibilities of intervention but also crucial existential choices that fit in the core of common and daily experiences, those regarding the different ways of giving and ending life, the care for the self and for others, the relationship with non-human living beings and with the entire biosphere. And this thick body of choices is once again connected to the theme of plurality. A plurality of paradigms of values outlining areas of highly conflictive tension concerning the different ways to intend the connection life-rules (Rodotà 2006) and so to define the concepts of liberty and autonomy, justice and solidarity.

Starting from the latter view we cannot deny that several bioethical issues were, and still are, triggered by the awareness of inequality and injustice, and thus by the urgent demand to work them out. Consequently, another source for the bioethical debate is the instance of equality and justice, which involves socio-political programs and visions of the world and of the destiny of peoples and individuals. The apparent injustice that the benefits related to life and health are far too often selectively available clearly plays a part, and the consequent denouncement acquires the form of "bioethical" theme, renewing the permanent need to root the reasons and enlarge the forms of responsibility. Responsibility is the duty that matches the honor to be a human being, thus it is not to be considered as a mere burden, but rather as a "response", according to its etymology, which stops us from being blind to the instances of reality and staying silent, that is to say incapable of making a stand. Indeed the "philosophical-moral" dimension of bioethics is marked by the willingness, which it is hoped becomes capacity, to respond and fight blindness and silence, that is strangeness to the processes concerning the human existence and its rich historical phenomenology.

As a matter of fact UNESCO deals with the complexity of the areas of intervention, the conceptual articulations and the political-institutional implications of the bioethical debate that we briefly mentioned previously. Over the years several initiatives have been promoted, at an international level, by this agency of the United Nations to favor the convergence towards indispensable principles meant to express the respect of human rights and fundamental freedoms.

The UNESCO Charter, signed on November 16, 1945, outlined the role of the organization underlining the pursuit of peace and security by reinforcing, through education, science and culture, "collaboration among the nations [. . .] in order to further universal respect for justice, for the rule of law and for the human rights and fundamental freedoms which are affirmed for the peoples of the world, without distinction of race, sex, language or religion, by the Charter of the United Nations" (Article 1).

Right from its genesis, then, UNESCO focuses on the strategic value of education, science and culture to promote international cooperation through relationships involving diversities that are productive of new ways of being and acting. The universality of this project is not intended as uniformity, where differences are cancelled, but as an ambitious search for a confrontation including multiple and diverse historical traditions and concepts of the world and of humankind. All too often plurality becomes a synonym of "incompatibility". The world we live in is undoubtedly plural so the acknowledgment of plurality seems unexceptionable, yet the further step, that is the achievement of compatibility, becomes a "narrow gate" : the "many" are "moral strangers" who fail to find a common language and, making their own idiom absolute, they end up by producing dimensions of incompatibility.

Article 12 clearly and strongly refers to cultural diversity and pluralism, which in the preamble are considered as "common patrimony of mankind": "the importance of cultural diversity and pluralism should be given due regard. However, such considerations are not to be invoked to infringe upon human dignity, human rights and fundamental freedoms, nor upon the principles set out in this Declaration, nor to limit their scope".

The acknowledgment of the value of cultural diversity is essential for a re-semantization of universality and to share the awareness that it is not Man but men and women who inhabit the Earth, and it is not one civilization that makes the world human but a multiplicity of cultures. This is the only way to overcome the sterile contraposition of conflictive visions chained to strict relativism or absolute metaphysics. In both cases we run the risk to assume the concept of culture as hypostasis without historicity, denying the cultures' dynamic liveliness which finds its unfolding in history itself. In this regard, the Declaration properly insists on the explicit reference to the respect for cultural diversity, which is not to result in the infringement upon human rights.

3. Meanings of dignity in the age of human rights and bioethics

Coming back to some of the elements mentioned in the introduction, it's necessary to reflect on the implications of framing bioethics within human rights, which is at the core of the UNESCO Declaration. It is a significant cultural and political option that contributes very effectively to determine the re-configuration of the "identity" of this subject of study, so that starting from an academic or mainly "ethical–philosophical" type of knowledge it acquires a more and more complex asset in a fully global scenario.

The axiological framework featuring in the Universal Declaration of Human Rights, explicitly transposed in the UNESCO Declaration, configures a shared kind of ethics already purified from the conflicts between philosophical and ideological positions. The Charter of 1948 features principles and values discussed, accepted and subscribed consensually by omitting any philosophical antithesis, and it uses the high and strongly authoritative language of human law. The successful osmosis between "ethical" and "juridical" vocabulary allows the Declaration to ambitiously re-propose sub-specie bioethical, so to say, the international law of human rights, which UNESCO undertook in 2003 and completed in 2005 after a laborious process of international consultation.

Therefore, within a framework firmly rooted in the vision of "the right to have rights" inalienably marking the human being, the notion of dignity is fast foregrounded: human rights and dignity provide the person with ethical and juridical consistency, giving new light to an old term. Indeed the word "person" is supposed to identify each and every one in the reality of their living, to give visibility to the "individuo innalzato a valore" (Bobbio 1944: 119), endowed with rights that in the second half of the 1900s started to represent their existential condition, the concrete historicity of belonging to different cultures and communities, acknowledging their body and needs of any order and degree.

The commitment of the UNESCO Declaration to formulate the principles of Bioethics starting from human rights implies the central role of dignity, no longer conceived in a prospective of hierarchical stratification but equally attributed to any person, granted and protected from any form of degradation, humiliation and violence. We have learned to comprehend the meanings of dignity after hard experience – destabilizing old beliefs – of abusing the living body of humankind. What was to be considered as concrete violation of dignity was defined over the decades that witnessed the rise of bioethics and bio-law, right from the Charters of dignity and the practices of law, from time to time "responsive" to actual or possible abuses. What history has shown "concavely", as calamities already occurred and possible risks, human rights and dignity – which is their matrix – have expressed "convexly", with words impossible to be deleted even if the plurality of the meanings they carry is still debatable. And over the last few years the possible normative use of dignity in bioethics has been the focus of a very lively international debate, in which the worst arguments claimed the "uselessness" of this notion due to its intrinsic vagueness (Macklin 2003: 1419). And the word "vagueness" disregards the polysemy and historical stratification of the values preserved by dignity. What is under debate is then a sort of "rhetoric" of dignity, all too often useful to conservative drifts, whose intent is to fix ex ante interdictions and bans on new applications of scientific and technological research. The conceptual themes of this debate are related to the sense that must be given to the relationship between moral and juridical normativity, and consequently to the multiple meanings that the notion of dignity may have in bioethics, which is a sector of deliberation made particularly complex by the necessary search for compatibility among the different positions emerging from the ethical reflection and the regulations provided by law at a regional, national and international level.

Analyzing the special trends in the international bio-juridical debate, Deryck Beyleveld and Roger Brownsword highlighted a significant change in the paradigm of dignity; in their 2001 monograph they distinguished between

> two conceptions of human dignity – 'human dignity as empowerment' and 'human dignity as constraint'. This distinction correlates broadly with the contrast between the background role typically assigned to human dignity in the founding international instruments of human right as against the foreground role assigned to it in the recent instruments that set the framework of modern bioscience. Where human dignity plays a background role, the governing conceptions human dignity as empowerment; where it plays a foreground role, the distinctive conceptions human dignity as constraint.
>
> (Beyleveld and Brownsword 2001: 11)

According to Beyleveld and Brownsword the concept of "human dignity as empowerment" was to be referred to the historical moment following the Second World War, when the notion of dignity played a pivotal role in the reinforcement of personal autonomy, marking the expansion of freedoms and rights and the recession of any form of external interference – even by the State – regarding individual choices. The concept of "human dignity as constraint", instead, referred to the notion developed within more recent documents of bioethics, where an objective sense of dignity becomes a source of duties and limits regarding individual autonomy.

The thesis expressed by Beyleveld and Brownsword met much criticism, particularly by Giorgio Resta who after a careful analysis considers it "infondata, o quanto meno inidonea ad offrire un valido ausilio ermeneutico per la comprensione dell'attuale contenuto operazionale della nozione di dignità" (Resta 2014: 11). Indeed Resta underlines the objective sense of dignity that belongs to the UNO Universal Declaration of 1948 and the major literature of postwar constitutionalism. Article 23 of the Universal Declaration of Human Rights affirms that "everyone who works has the right to just and favorable remuneration ensuring for himself and his family an existence worthy of human dignity, and supplemented, if necessary, by other means of social protection". This aspect of dignity as a limit to private autonomy appears also in Article 36 of the Italian Constitution. However, as Resta underlines, the meaning of dignity as empowerment is still effective in more recent international documents concerning bioethical issues. This is the case, for instance, of the UNESCO Declaration on Human Genome, where dignity is mentioned as the key to restrain reductionist drifts and avoid limiting individual identities according to their genetic features, so "to respect their uniqueness and diversity".

Just a brief mention of these different positions can lead to the comprehension of the general guidelines of a debate that has grown in the international scenario and has surely not come to an end yet.

A reference for such debate is the editorial by Ruth Macklin published in the *British Medical Journal* in 2003, in which the author states:

> Appeals to human dignity populate the landscape of medical ethics. Claims that some feature of medical research or practice violates or threatens human dignity abound, often in connection with developments in genetics or reproductive technology. But are such charges coherent? Is dignity a useful concept for an ethical analysis of medical

activities? A close inspection of leading examples shows that appeals to dignity are either vague restatements of other, more precise, notions or mere slogans that add nothing to an understanding of the topic.

(Macklin 2003: 1419)

Macklin's target is especially the document by the US Council on Bioethics appointed by President George Bush, in which a remarkable attention is paid to the concept of dignity. Yet, according to Macklin, "the report contains no analysis of dignity or how it relates to ethical principles such as respect for persons. In the absence of criteria that can enable us to know just when dignity is violated, the concept remains hopelessly vague" (Macklin 2003: 1420). The US President's Council on Bioethics tried to reply in 2008 with a ponderous volume containing 19 essays and commentaries by influential scholars invited to discuss the questions opening the introduction by Adam Schulman:

Human dignity – is it a useful concept in bioethics, one that sheds important light on the whole range of bioethical issues, from embryo research and assisted reproduction, to biomedical enhancement, to care of the disabled and the dying? Or is it, on the contrary, a useless concept – at best a vague substitute for other, more precise notions, at worst a mere slogan that camouflages unconvincing arguments and unarticulated biases?

(Schulman 2008: 3)

The answers given by experts such as Daniel C. Dennett, Patricia S. Churchland, Holmes Rolston III, Nick Bostrom, Martha Nussbaum, and Edmund D. Pellegrino deal with the theme of dignity from different points of view and do not allow to get to a unitary definitional conclusion. But such difficulty in sharing a definition should not lead to cancel the word "dignity" in the bioethical and bio-law vocabulary; it should rather urge the comprehension of its semantic complexity which cannot be limited to merely instrumental and ideological uses, and then put forward the need to a wider interpretation in order to catch both the historical and philosophical aspect of this concept and thus avoid any distortion or ambiguity.

The absence of a definition of dignity is maybe a felix culpa of the UNESCO Declaration as well as of the organism's previous documents. The intrinsic univocal vocation of a definition would frame the polysemy of the word – a polysemy connected to the historical experience – in a rigid and paralyzing scheme. If "dignity" is one of the great terms of the human lexicon, if it acquires sense and value in human lifetime, then it cannot and must not be defined in some essentialist key; on the contrary, it has to be built up starting from history, from its philosophical conceptualization, from its belonging to the language of law – ranging from constitutional law to that of human rights – and from the different types of practice in which it played a part, first of all the jurisprudential one.

Inscribing the debate within a wider historical and philosophical frame, Stefano Rodotà refers to a "revolution of dignity" wellspring of a figure of human being, an anthropology of the homo dignus whose nucleus is the person different from any abstract subjectivity and thus intended "come la categoria che meglio permette di dare evidenza alla vita individuale e alla sua immersione nelle relazioni sociali" (Rodotà 2012: 183), up to the vertices of the "constitutionalized person".

Taken away from the claim of tradition, whose limits and possible evolutive and prospective interpretations are acknowledged, dignity becomes the keystone of a unitary logic of rights which indissolubly connects it to freedom, equality and justice. "La dignità" – Rodotà writes – "non è un diritto fondamentale tra gli altri, né una supernorma. Seguendo la storia della sua vicenda giuridica, ci avvediamo che essa è venuta a integrare principi fondamentali già consolidati – libertà, eguaglianza, solidarietà – , facendo corpo con essi e imponendone una reinterpretazione in una logica di indivisibilità" (Rodotà 2012: 199).

4. The novum of vulnerability

In the self-conscience expressed by humankind in the great Declarations of the 1900s, dignity, freedom and equality belong to one another. At least from a conceptual point of view they are intertwined by means of a link that cannot be weakened by defining freedom, in the age of bioethics, mainly as autonomy and self-determination.

The UNESCO Declaration undertakes this mutual belonging and sets the conditions to avoid conflicts by insisting on the relevance of consent as an instrument to achieve autonomy. An informed consent allows the "knowledge" of the patient, surely different from the doctor's one yet equally essential, the competence that guides the life of every individual. Therefore consent grants everybody, even in a condition of fragility, the chance to practice freedom and capacity of judgment, to evaluate their own interests, measuring them to their own ethics and personal lifestyles.

The informed consent allows to fulfill the duty of respecting one's own autonomy and self-determination and thus freedom and capacity to decide by oneself, whatever the territory the self encompasses, since the human self is not an atom but a crossover of relationships.

By the tight link connecting autonomy and responsibility, and therefore one's own subjectivity and duty towards the others, the UNESCO Declaration protects the principle of autonomy from solipsistic trends, inscribing relatedness within the core of human identity. Moreover, the document explicitly mentions "social responsibility" (Article 14) in the inclusive context of communal life with its needs and demands of solidarity.

Solidarity is one of the major words of the UNESCO Declaration, both as a duty and a right and above all connective between individuals and peoples, evoked in the diversity of their conditions in Article 24: "[. . .] solidarity between and among States, as well as individuals, families, groups and communities, with special regard for those rendered vulnerable by disease or disability or other personal, societal or environmental conditions and those with the most limited resources".

Proposing explicitly as a feature of the human being, the supportive relationship among generations as well as the interconnection "between human beings and other forms of life", with the biosphere as a whole, the Declaration makes "responsibility" a key word by extending its scope. Mentioning responsibility, the Declaration pays special attention to the acknowledgment of the sense of vulnerability, referred to peoples and human beings exposed to the vulnus of life and death, thus structurally "fragile", regardless of the ways in which fragility appears in the course of life.

And with the aim to cope with vulnerability science has provided remedies by extending our freedom in the attempt to heal the wounds deriving from fragility. And also ethics, with its own reasons within the frame of the pursuit of social good, urges the full consideration of vulnerability; moreover, law and politics claim to defend vulnerability according to the principle of "the right to have rights". If vulnerability is not an "accidental" dimension but a defining trait of the human condition, if its principle is a way to comprehend humankind and design their individual and collective daily existence, then it is the premise to consider the various forms of fragility in terms that exclude the "exception", the "negative privilege" and include them in the commonly shared concept of "citizenship".

The same autonomy can be defined according to a scale of levels of exercise which includes the weaker categories and pays attention to the lowest conditions in which it cannot be implemented at all. In the latter case, as dignity is inalienably connected to the identity of an individual even in a situation of impossible exercise of freedom, the "incapable" will not be exposed to the terrible risk of indignity. A risk that history has tragically experienced and that the clear language of law is supposed to contrast, making the interpretations from the past of humankind ineffective and meaningless. The informed consent is the means by which, in medicine and experimental research, autonomy is made inviolable. Article 7 of the UNESCO Declaration outlines the wide background where such consent has to be formulated at all conditions: "[. . .] the person concerned should be involved to the greatest extent possible in the decision-making process of consent, as well as that of withdrawing consent".

Therefore the principle of vulnerability becomes the "remedial" for the issue of autonomy once conceived as possible only for "capable" subjects. Principles and rights protect from "becoming unworthy" due to the loss of autonomy. As a result, the formulation of the principle of autonomy involves the commitment to improve the conditions that make it possible, and so it has performative potentialities.

The attention paid "to the distinctness of each individual life" (Nussbaum 2006: 282), to the singleness of the person that is always characterized by ties of dependence and interdependence, is the keystone of a theoretical reflection which promotes a close connection between human dignity and the needs characterizing our concrete existential conditions; all human beings have to deal with them in order to be able to outline the field of their freedom, and thus independently make their own project of life. In this same direction, particularly significant are Martha Nussbaum's ideas; staying within the concept of capability approach already outlined by Amartya Sen and adopting a newly Aristotelian point of view, she investigates "the prerequisites for living a life that is fully human rather than subhuman, a life worthy of the dignity of the human being" (Nussbaum 2006: 278). Nussbaum underlines that "need and capacity, rationality and animality are thoroughly interwoven, and that the dignity of the human being is the dignity of a needy enmattered being" (Nussbaum 2006: 278). Focusing on what people are actually able to do and to be, the approach developed by the scholar leads to a revision of the theories of justice, which implies the urgency to face the wide range of vulnerabilities related to the human condition including all the inequalities, especially those connected to gender which Nussbaum considers as the true test for the most relevant political, economic and cultural choices, both national and international. And if "people are entitled not only to mere life, but to a life compatible with human dignity, and this entitlement means that the relevant goods must be available at a sufficiently high level" (Nussbaum 2006: 292), then "it is the equal dignity of human beings that demands recognition", as it is related to the "idea of the social bases of self-respect and non-humiliation" (Nussbaum 2006: 292).

5. The commitment to education and training

The policy of conceptual elaboration and sharing of "universal principles based on shared ethical values to guide scientific and technological development and social transformation", which we have so far mentioned, is reinforced by the promotion of bioethics education, training and information that the Declaration explicitly considers as an operative commitment. Indeed Article 23 states that

> in order to promote the principles set out in this Declaration and to achieve a better understanding of the ethical implications of scientific and technological developments,

in particular for young people, states should endeavor to foster bioethics education and training at all levels as well as to encourage information and knowledge dissemination programs about bioethics.

This position, which finds previous significance in the "Explanatory Report to the Oviedo Convention" and in the "Universal Declaration on the Human Genome and Human Rights" adopted unanimously at UNESCO's Twenty-ninth General Conference on November 11, 1997, is symptomatic of a new emphasis placed on information and education as an essential assumption for the acquisition of individual awareness as well as for the democratic involvement in public debate. Thus, taking into account a general social difficulty in metabolizing the innovative processes that upset life and the coordinates of an anthropology rooted in natural laws considered as insurmountable until a few decades ago, the issue of educating for an authentic comprehension of the conceptual and practical changes caused by the development of science and technology appears as an undelayable responsibility that requires commitment not only at a cultural level but also at an institutional one.

This policy is precisely confirmed also in other documents published by UNESCO in the years that followed the Declaration, such as the Draft Report on Social Responsibility and Health released by the UNESCO International Bioethics Committee in 2009; it analyzes the impact of health literacy, that is the individuals' capacity to comprehend and manage the key factors to their own health, and it underlines the role of governments in promoting opportunities for bioethics education, always functional to a proper participation in the public debate.

Bioethics training is then to be considered as a decisive joint of the "knowledge society" which represents the primary lever for the construction of new dimensions of citizenship based on the expansion of a global space of rights.

Among the activities promoted by UNESCO it's possible to point out important initiatives dedicated to the relationship between bioethics and education. Starting from 2004 the Division "Ethics of Science and Technology" has planned an Ethics Education Program (EEP) arranged with several directions of intervention.

The first one regards the implementation of a Global Ethics Observatory that allows to know and compare programs of teaching and national legislations. The second pointer concerns the organization of training courses for teachers, and the third area of activity involves the implementation of a Core Curriculum for the teaching of Bioethics, which took form in 2008 with the publication of a Syllabus displaying the essential elements of the ethics education program. The Introduction soon makes clear that the goal of the Core Curriculum is not to impose "a particular model or specific view of bioethics, but articulates ethical principles that are shared by scientific experts, policy-makers and health professionals from various countries with different cultural, historical and religious backgrounds".

The Syllabus is composed of seventeen units, whose aim is to promote the capacity to identify the ethical issues related to the progress of the sciences of life and also to justify the choices through rational argumentations, using as a reference the principles expressed in the Universal Declaration on Bioethics and Human Rights of 2005. Unit 3, for instance, highlights very significantly the pertinence of the concept of human dignity in the context of bioethical investigation: the process is arranged in several steps that suggest – starting from an insight into the notion of dignity in the history of ideas, with a special focus on Kant's conception, and in the great Bills of Rights like the Universal Declaration of Human Rights of 1948 – a dynamic and lay view of the acknowledgment of personal dignity in all its manifestations in private and public life. Unit 8 is rich in interesting scopes too, as it deals with the issues of respect for human vulnerability and personal integrity: assuming a wide point of view on the semantics of vulnerability, considered not only in its physical and biological aspects but also in its social ones, the model proposes the interpretation of the powers of medicine as an instrument to cope with those situations of suffering and pain that belong to the human condition, whose main feature is the constant exposure to vulnerability.

Besides the Syllabus, UNESCO has promoted the publication of a series of volumes suggesting a different approach based on the study of cases. The intent is to adopt a strategy that aims at involving young people in the discussion of new dilemmatic situations connected to the progress of the sciences of life.

These contributions want to propose a reflection on the methodology that may allow the transition, already advocated by the ONG-UNESCO Comité de Liaison in 2007, from the teaching of bioethics to the education of bioethics, that is to say from the practice based on the transmission of knowledge to the research of new modalities of training mediation suitable for endowing the new generations with the capacity to question themselves knowingly on the implications of science and technology, reaching an inter-disciplinary vision, inclusive and open to a confrontation with values and traditions from different cultures.

In this respect, UNESCO's commitment to education is strongly based on the philosophical implant of values of the Declaration of 2005, and it even contributes to enlarge its scope. Such scope, within a unitary frame of extension of rights and acknowledgment of freedom and responsibility of the scientific research, includes the promotion of practices of inter-generational educational dialogue, which are the ineradicable assumptions to outline forms of coexistence whose foundation is respect and solidarity.

Bibliography

Ambroselli, Claire 2005, Eugenisme, crime contre l'humanité et droits universelles, *Revue d'histoire de la Shoah*, 183: 457–503.

Arendt, Hannah 1979, *The Origins of Totalitarianism*, 3rd edition, San Diego, New York and London: Harcourt Brace & Company.

Arendt, Hannah 1998, *The Human Condition*, 2nd edition, Chicago and London: The University of Chicago Press.

Barcaro, Rosangela, Becchi, Paolo, Donadoni, Paolo 2008, *Prospettive bioetiche di fine vita. La morte cerebrale e il trapianto di organi*, Milano: FrancoAngeli.

Becchi, Paolo 2013, *Il principio dignità umana*, Brescia: Morcelliana.

Beecher, Henry, Adams, Raymond D., Barger, Clifford A., Curran, William J., Denny-Brown, Derek, Farnsworth, Dana L., Folch-Pi, Jordi, Mendelsohn, Everett I., Merrill, John P., Murray, Hoseph, Potter, Ralph, Schwab, Robert, Sweet, William 1968, A Definition of Irreversible Coma: Report of the ad hoc Committee of the Harvard Medical School to Examine the Definition of Brain Death, *Journal of the American Medical Association*, 205(6): 337–340.

Beyleveld, Deryck, Brownsword, Roger 2001, *Human Dignity in Bioethics and Biolaw*, Oxford: Oxford University Press.

Bobbio, Norberto 1944, *La filosofia del decadentismo*, Torino: Chiantore.

Böckönferde, Ernst-Wolfgang 2010, *Dignità umana e bioetica*, Brescia: Morcelliana.

Bostrom, Nick 2005, A History of Trans-Humanist Thought, *Journal of Evolution and Technology*, 14(1), www.nickbostrom.com.

Braidotti, Rosi 2013, *The Post-Human*, Cambridge: Polity Press.

Buchanan, Allen 2011, *Better than Human: The Promise and the Perils of Enhancing Ourselves*, Oxford: Oxford University Press.

Buchanan, Allen, Brock, Dan W., Daniels, Norman, Wikler, Daniel 2001, *From Chance to Choice: Genetics and Justice*, New York: Cambridge University Press.

Capograssi, Giuseppe 1959, La Dichiarazione universale dei diritti dell'uomo. In Id. *Opere*, 7 vols, vol. 5 (pp. 35–50), Capograssi, Giuseppe (ed.), Milano: Giuffré.

Cranford, Ronald E. 2004, Death, Definition and Determination of, I. Criteria for Death, in *Encyclopedia of Bioethics*, 3rd edition, vol. 2 (pp. 602–608) Post, Stephen G., ed., New York: Macmillan Reference.

Curi, Umberto 2015, *La porta stretta*, Torino: Bollati Boringhieri.

D'Antuono, Emilia 2014, Le frontiere mobili del possibile. Tra scienza e ethos. In *Frontiere mobili. Implicazioni etiche della ricerca biotecnologica* (pp. 195–211) Chieffi, Lorenzo, ed., Milano: Mimesis.

d'Avack, Lorenzo 2015, *Il potere sul corpo. Limiti etici e giuridici*, Torino: Giappichelli.

de Ceglia, Francesco Paolo (ed.) 2014, *Storia della definizione di morte*, Milano: Franco Angeli.

Di Ciommo, Mario 2010, *Dignità umana e Stato costituzionale: la dignità umana nel costituzionalismo europeo, nella Costituzione italiana e nelle giurisprudenze europee*, Firenze: Passigli.

Emanuel, Ezekiel J., Grady, Christine C., Crouch, Robert A., Lie, Reidar K., Miller, Franklin G., Wendler, David D. (eds.) 2008, *The Oxford Textbook of Clinical Research Ethics*, Oxford: Oxford University Press.

Engelhardt, Hugo Tristam Jr., 1996. *The Foundations of Bioethics*, 2nd edition, New York: Oxford University Press.

Flick, Giovanni Maria 2015, *Elogio della dignità. Se non ora quando?*, Roma: Libreria Editrice Vaticana.

Fukuyama, Francis 2002, *Our Post-Human Future: Consequences of the Biotechnology Revolution*, New York: Picador, Farrar, Straus & Giroux.

Furlan, Enrico (ed.) 2009, *Bioetica e dignità umana: interpretazioni a confronto a partire dalla Convenzione di Oviedo*, Milano: Franco Angeli.

Gervais, Karen G. 2004, Death, Definition and Determination of, III. Philosophical and Theological Perspectives, in *Encyclopedia of Bioethics*, 3rd edition, vol. 2 (pp. 615–625). Post, Stephen G., ed., New York: Macmillan Reference.

Habermas, Jürgen 2001, *Die Zukunft der menschlichen Natur: auf dem Weg zu einer liberalen Eugenik?*, Frankfurt am Main: Suhrkamp.

Harris, John 2007, *Enhancing Evolution: The Ethical Case for Making Better People*, Princeton: Princeton University Press.

Hottois, Gilbert 2009a, La techno-science met-elle en danger la diversité culturelle?, *Journal International de Bioéthique*, 20(1–2): 17–30.

Hottois, Gilbert 2009b, *Dignité et diversité des hommes*, Paris: Librairie philosophique J.Vrin.

Hottois, Gilbert 2012, Quelques remarques au sujet de la notion de techno-science, *Al-Mukhatabat, Revue Philosophique électronique pour la Logique, l'Epistémologie et la Pensée Analytique*, 2: 24–28, <http://almukhatabatjournal.l.a.f.unblog.fr/files/2012/11/gilbert-hottois.pdf>.

Hottois, Gilbert 2014, *Le transhumanisme est-il un humanisme?*, Bruxelles: Académie Royale de Belgique.

Hottois, Gilbert, Missa, Jean-Noël, Perbal, Laurence (dir.) 2015, *Encyclopédie du trans/posthumanisme. L'humain et ses préfixes*, Paris: Vrin.

Hughes, James 2004, *Citizen Cyborg: Why Democratic Societies Must Respond to the Redesigned Human of the Future*, Cambridge: Westview Press.

Jonas, Hans 1979, *Das Prinzip Verantwortung: Versuch einer Ethik für die technologische Zivilisation*, Frankfurt am Main: Insel-Verlag.

Jonas, Hans 1985, *Technik, Medizin und Ethik. Zur Praxis des Prinzips Verantwortung*, Frankfurt am Main: Insel-Verlag.

Lenoir, Noëlle, Mathieu, Bernard 1998, *Les normes internationales de la bioéthique*, Paris: PUF.

Levi, Primo 2015, The Drowned and the Saved, trans. Michael F. Moore, in *The Complete Works of Primo Levi*, vol. 3 (pp. 3590–3591) Goldstein, Ann, ed., New York and London: Liveright Publishing Corporation.

Lévinas, Emmanuel 1974, *Autrement qu'être ou au-delà de l'essence*, La Haye: Martinus Nijhoff.

Macklin, Ruth 2003, Dignity Is a Useless Concept: It Means No More than Respect for Persons or Their Autonomy, *British Medical Journal*, 327: 1419–1420.

Maritain, Jacques 1990, Sur la philosophie des droits de l'homme, in *Œuvres complètes*, 15 voll., vol. 9 pp. 1906 – 1920), Fribourg and Paris: Editions Universitaires.

Miller, Franklin G., Truog, Robert D. (eds.) 2012, *Death, Dying, and Organ Transplantation: Reconstructing Medical Ethics at the End of Life*, Oxford: Oxford University Press.

Munson, Ronald 2007, Organ Transplantation, in *The Oxford Handbook of Bioethics* (pp. 211–239) Steinbock, Bonnie, ed., Oxford: Oxford University Press.

Nussbaum, Martha C. 2000, *Women and Human Development: The Capabilities Approach*, Cambridge and New York: Cambridge University Press.

Nussbaum, Martha C. 2006, *Frontiers of Justice: Disability, Nationality, Species Membership*, Cambridge and London: The Belknap Press of Harvard University Press.

Resta, Giorgio 2014, *Dignità, persone, mercati*, Torino: Giappichelli.

Rodotà, Stefano 2006, *La vita e le regole. Tra diritto e non diritto*, Milano: Feltrinelli.

Rodotà, Stefano 2010, Antropologia dell'homo dignus, Lectio doctoralis, in *Conferimento della Laurea honoris causa in Scienze della politica a Stefano Rodotà*, Macerata: Biemmegraf. Civilistica.com 2013(2): 1–13.

Rodotà, Stefano 2012, *Il diritto di avere diritti*. Roma-Bari: Laterza.

Rosen, Michael 2012, *Dignity: Its History and Meaning*, Harvard: Harvard University Press.

Sandel, Michael J. 2007, *The Case against Perfection: Ethics in the Age of Genetic Engineering*, Cambridge and London: Belknap Press, Harvard University Press.

Savulescu, Julian, Bostrom, Nick (eds.) 2009, *Human Enhancement*, Oxford: Oxford University Press.

Schulman, Adam 2008, Bioethics and the Question of Human Dignity. In President's Council on Bioethics. Human Dignity and Bioethics, Washington.

Sen, Amartya 1985, *Commodities and Capabilities*, Amsterdam: North-Holland.

Sen, Amartya 1993, Capability and Well-Being, in *The Quality of Life* (pp. 30–53) Nussbaum, Martha, Sen, Amartya, eds., Oxford: Clarendon Press.

Sen, Amartya 1999, *Development as Freedom*, New York: Knopf.

Sennett, Richard 2003, *Respect in a World of Inequality*, New York: W. W. Norton & Company.

ten Have, Henk 2006, The Activities of UNESCO in the Area of Ethics, *Kennedy Institute of Ethics Journal*, 16(4): 333–351.

ten Have, Henk (ed.) 2015, *Bioethics Education in a Global Perspective: Challenges in Global Bioethics*, Dordrecht, Heidelberg, New York, and London: Springer.

Vincenti, Umberto 2009, *Diritti e dignità umana*, Roma-Bari: Laterza.

Youngner, Stuart J. 2007, The Definition of Death, in *The Oxford Handbook of Bioethics* (pp. 285–303) Steinbock, Bonnie, ed., Oxford: Oxford University Press.

Youngner, Stuart J., Arnold, Robert M., Schapiro, Renie (eds.) 1999, *The Definition of Death. Contemporary Controversies*, Baltimore and London: The Johns Hopkins University Press.

Zatti, Paolo 2009, *Maschere del diritto volti della vita*, Milano: Giuffrè.

Part II

The contribution of the Universal Declaration on Bioethics and Human Rights to the construction of a shared bioethics

VI The Universal Declaration on Bioethics and Human Rights

Towards intercultural bioethics and global justice: the case of transnational research

Laura Palazzani

Towards intercultural bioethics

Each culture has perceived and interpreted in its own way the importance and role of bioethics in theoretical and applied fields. There is a huge variety of bioethical questions and responses within different cultures: the diverse levels of development of scientific knowledge and technological applications in biomedicine raise different issues and as such establish a different priority in the urgency of solutions, given the different socio-economic and political-juridical contexts.

The heterogeneous theoretical and practical settings of the many cultures (the beliefs, conceptions of philosophy and religion, values and principles, traditions, customs and habits), but also the particular way cultures relate to techno-scientific innovation, as well as the specificity of the political, legal and social context, are certainly factors in diversification. In this sense, even in bioethics, there is an increasingly important role of the description and interpretation of the cultural context in which diverse bioethical theories and issues originate and develop.

In-depth study in this field is becoming more necessary because of the new bioethical issues arising from the coexistence of different ethnic groups in the same territory and because of the issues emerging from the relationship between different (even geographically distant) cultures. It is becoming increasingly evident in our multiethnic societies that the recipients of care and application of the latest scientific technologies, but also the doctors and health professionals applying them, are individuals who cohabit in the same social reality despite their often having different cultural roots; different conceptions of life, birth, suffering, death, health and illness. Moreover, it is clear that solutions to bioethical problems within a culture often have implications, be they immediate or future, direct or indirect, or external in relation to the specific historical and social conditions from which they originate: i.e. pandemics, genetic manipulations, international multi-centric trials, environmental issues. There is an ever-growing danger that bioethics directs its attention only internally to the problems of one's own culture without looking externally to the problems of other cultures, especially those in objectively disadvantaged conditions.

There is a need for a macro-bioethics or global bioethics, broadened in space, across cultures, countries, continents and in time, between distant and future generations. A global bioethics includes necessarily an intercultural dialogue (European Commission, 2010).

This issue is central in bioethical reflection and has given rise to different theories of the relationships between different cultures.

The ethnocentric paradigm considers one's culture as being superior to all others in a hierarchical view. This theory considers in an exclusive manner one cultural perspective as "the" predominant bioethics, with the consequent imposition over other cultures. Bio-ethical ethnocentrism expresses itself in the so-called bioethical imperialism or bioethical colonialism or, expressed in a softer way, bioethical paternalism. Ethnocentric bioethics presupposes the superiority of one's cultural values, considering unnecessary any comparison with those of other cultures, and imposes itself on weaker cultures, deemed inferior.

It is a perspective that proposes the "model of assimilation", requiring that those belonging to other cultures adapt and adjust to the bioethics of the "mainstream" culture and the "model of subordination", with consequent (possible) exploitation. Ethnocentric bioethics is considered unacceptable as it proposes in an arbitrary manner a claim for superiority, with an attitude of unjustified intolerant "arrogance".

Multicultural paradigm, in the relativist philosophical prospective, undermines ethnocentric bioethics considering the bioethics of each culture as placed on the same level with regard to the bioethics of any culture, all cultures being considered equivalent. It is the theory suggesting an attitude of tolerance, interpreted as passive and indulgent acceptance of all manifested bioethics in different cultures. It is the relativist prospective which, from the viewpoint of the juxtaposition of multiple and diverse bioethics, believes the plurality of bioethics to be irreducible to unity. In this sense multicultural bioethics considers the search for common values futile and undesirable, considering plurality better than unity as an expression of richness and originality. Unity, instead, would be oppressive and suffocating (Baker, 1998).

It is a perspective that contrasts assimilation with the "model of separation": each culture is represented and perceived as a "closed" world, which internally affirms its own values and preserves its own traditions and externally tolerates any other bioethical culture.

The principle of multicultural equivalence, despite neutral and uncritical acceptance of every culture, is unable to avoid cultural conflict, historically evident, among cultures which are opposing and incompatible, contextual and simultaneous, allowing the stronger culture to prevail over the weaker one. Equivalence can lead to self-closure of each culture in itself, resulting in incommunicability.

In opposition to the ethnocentric and multicultural paradigms in bioethics, the intercultural perspectives try to overcome the hierarchy among cultures and at the same time their equivalence. This intermediate proposal is possible in the human rights framework, considering the human dignity as the minimum common value. According to this perspective, equality must ensure that all human beings, in the universal sense, regardless of cultural belonging, have dignity. Affirmation of the principle of equality does not mean to ignore or set aside or, worse still, suppress differences. Affirming equality means assuming the differences. Equality is the precondition for "recognition" of the "differences", which is not intended only as acknowledgement of diversity, but as significant interaction between human beings.

In this perspective, the role of bioethics is not to impose a view as superior denying and suppressing the others, nor is it to legitimize uncritically every request from each cultural group. The structural reference to human rights acquires undoubted priority in ethno-bioethics. The theory of human rights is the "unit of meaning" in relation to which compatibility between the rights of different cultures can be implemented moving towards interculturalism. Pursuing this direction avoids some negative paths: such as those that demand egalitarianism in assimilation, believing that all humans of diverse cultures should conform and comply with the dominant culture or those that demand differentiation in separation, considering that the individuals of different ethnic groups should be guaranteed by the broadest possible conservation of what makes them different.

The importance of intercultural bioethics consists in the critical search for continuous mediation and integration of human rights and the specific needs of diverse cultures, in an attempt to prevent the abuse and affirming the relational approach of diversity in equality. Consequently, intercultural dialogue in bioethics would be neither hostile competition nor passive acceptance of a pragmatic compromise, but rather a constructive search for integration.

The Universal Declaration on Bioethics and Human Rights gives a specific contribution to the elaboration of an intercultural bioethics both on the level of theory and of practice.

The document outlines a "universal framework" or the minimum principles considered universally essential for bioethics, identified in the promotion of "respect for human dignity", "respect for the life of human beings, and fundamental freedoms, consistent with international human rights law". The principles set out express the recognition of the duty to respect human dignity and human rights in a plural cultural setting, given the new possibilities opened up by techno-science in the biomedical field. Article 3 states that: "Human dignity, human rights and fundamental freedoms are to be fully respected"; "The interests and welfare of the individual should have priority over the sole interest of science or society".

The document recognizes human dignity as a meta-cultural or transcultural value, stressing the fact that "no individual or group should be discriminated against or stigmatized on any grounds, in violation of human dignity, human rights and fundamental freedoms" (. . .). Article 12 on "Respect for cultural diversity and pluralism" states: "The importance of cultural diversity and pluralism should be given due regard. However, such considerations are not to be invoked to infringe upon human dignity, human rights and fundamental freedoms, nor upon the principles set out in this Declaration, nor to limit their scope".

Towards global justice and international health cooperation

This perspective of intercultural bioethics leads to comprehension of the meaning of global justice and international health cooperation, on a theoretical and applied level.

In a globalized world, especially in the field of health, the gap between the so-called advanced/developed and developing countries is becoming increasingly evident (Falk, 1999). On the one hand, Western countries are steadily thrust towards further improvements in scientific and technological progress directed at guaranteeing even better protection of health in the health system, in a quantitative and qualitative sense: i.e. the recent developments in reproductive technologies, genetic discoveries, experimental intervention in the desire to cure "at all costs", the enhancement frontiers, that is, medical intervention not for therapeutic purposes but only for amelioration/improvement. On the other hand, developing countries are unable to guarantee the survival of individuals (consider the problem of hunger and poverty), in circumstances where technology is scarce and rudimental, and there is a lack of basic health care.

In-depth analysis on the distribution of goods in the world (in relation to economic and health resources) reveals that the world situation is characterized by the existence of sufficient quantities to address the basic demands of everyone with regard to survival and health, but that they are unevenly distributed. The amount of food and medicine in the world is enough to meet global needs, but not everyone has access: only those with adequate economic resources or those living in places where the juridical system makes feasible the possibility of receiving assistance and care even in the absence of economic resources.

Therefore, there is the perception of "global injustice" and the need to expand the concept of "health" and the "right to safeguard health" in a dimension that goes beyond localism. Intercultural bioethics highlights the fact that it is essential to broaden our vision beyond political and cultural boundaries in order to guarantee primary access to basic health care for everyone regardless of cultural belonging. The goal can only be a common intercultural one: guaranteeing equity in health on a global scale. It is a perceived problem in concrete terms and it has been the object of reflection within specific theories. Although the sensibility detected regarding unequal distribution of resources in space (in synchronic terms) is inferior compared with the sensibility addressed to future generations for "sustainable development" (in diachronic terms). There is a widespread conviction that indiscriminate exploitation of natural resources must be avoided for intergenerational justice; less widespread is the idea of equal distribution of available goods with a view to global justice.

The perspective of intercultural bioethics encompasses justification of significant genuine instances of global justice. The Declaration of Bioethics and Human Rights sets out in Article 10 "Equality, justice and equity": "The fundamental equality of all human beings in dignity and rights is to be respected so that they are treated justly and equitably". Article 11 on non-discrimination and non-stigmatization affirms: "No individual or group should be discriminated against or stigmatized on any grounds, in violation of human dignity, human rights and fundamental freedoms". Article 14 on social responsibility and health underlines the relevance of access to quality health care and essential medicines, especially for the health of women and children, because health is essential to life itself and must be considered to be a social and human good and elimination of the marginalization and the exclusion of persons on the basis of any grounds.

In this perspective, access to health resources should not depend on the free market or the calculation of social productivity, but be guaranteed to everyone with a view to substantial, intercultural, international and global justice. What should be placed at the centre of the argument is not the libertarian principle of individual self-determination or the utilitarian principle of social convenience, but the dignity of each human being 'beyond' autonomy and utility. Society is called upon to take responsibility for unjust inequalities, giving subsistence to the right of individuals and disadvantaged peoples to assert claims (as moral rights) and a social duty to solidarity, cooperation and altruism (Article 13).

It is not a task of intercultural bioethics of developing strategies to achieve global health justice: it is the task of epidemiology, health statistics and bio-economy, together with sociology and cultural anthropology, to address the complex problem of the "rationalization" of health care costs, balancing the different needs of different cultures by quantifying needs and designing measures to meet basic needs. Whereas what is incumbent on bioethicists, first and foremost, is the commitment to shed light on the fact there are people in the world, cultural areas and groups of particularly vulnerable individuals (within different cultures), in objectively disadvantaged conditions, their being given assistance and help is a fundamental public duty and it is not lawful to operate against them any rational calculation of cost saving according to an approach of imposition of strength and exploitation for selfish advantage; secondly it is a fundamental bioethical requirement to prevent the allocation of health resources from violating some fundamental bioethical principles, the first among them being equality (there can be no social or cultural discrimination of any kind among the sick benefitting from health resources); thirdly the promotion of training allowing those who are disadvantaged not only to receive adequate assistance with relation to needs, but to acquire the skills to be able to collaborate, participate and intervene even actively (and not only passively) to improve their conditions.

The common goal outlined in intercultural bioethics is clearly the effort to achieve a progressive alignment of standards in health and welfare in all countries of the world, as well as suitably taking into account the economic interests that support investment strategies in their public and private aspects, so that the exclusive profit-oriented rationale, strict but blind in terms of values, will not be the only guiding strategy, but rather a wise mixture of market mechanisms and substantive justice. This is the only perspective in which to establish the principle of global health cooperation on humanitarian grounds.

Transnational research: the case of clinical trials in developing countries

The globalization of research would increase the conditions of justice and equality in the distribution of drugs (Benatar and Singer 2000; Emanuel et al. 2004; Hyder et al. 2004; Wendler et al. 2004; Petryna 2005; Glickman et al. 2009). Although the globalization of clinical studies hides, often, the objective of "outsourcing" the experimentation, in order to reduce costs, simplify and accelerate procedures. The reference is to the experimentations that involve those populations that are particularly "vulnerable" mainly because of economic underdevelopment that slows down the progress of science and technology or, even if economically developed, unaware of ethical issues (Angell 1997, 2006; Varmus and Satcher 1997; Koski and Nightingale 2001; Shapiro and Meslin 2001; Hawkins and Emanuel 2008; Lorenzo et al. 2010). These conditions may expose some populations to a risk of exploitation for scientific interests, which may hide commercial interests (Pavone 2016: 76). It may be considered as a form of bioethical "colonialism" and "imperialism", unfair exploitation and manipulation due to the differences in scientific-technological knowledge and socio-economic and cultural inequalities (Benatar 1998).

The Universal Declaration on Bioethics and Human Rights expresses the general framework of reflection with references to human dignity (Article 3), the direct and indirect benefits for patients participating in the research (Article 4), informed consent (Article 6), respect for human vulnerability and personal integrity (Article 8), equality, justice and equity (Article 10), non-discrimination (Article 11), respect for cultural diversity (Article 12), solidarity and cooperation (Article 13), social responsibility and health as a fundamental human right (Article 14), international cooperation (Article 24), promoting the international dissemination of scientific information, freedom of movement and sharing of scientific and technological knowledge.

The Declaration recalls the general ethical principles of experimentation on human subjects – recognized in international documents, affirming that they should be applicable everywhere, without making a distinction between more or less developed countries, avoiding unequal treatment and recognizing the universal justice. This does not mean accepting a "double standard" of ethics (Macklin 2004): on the contrary, it means reiterating that the ethical standard should be "unique" as concerns principles. Trials in developing countries must meet the same ethical standards of developed countries (Article 21 b).

The general ethical standards which must be considered mandatory, as substantive ethical requirements for clinical trials on the international level are the protection of all human subjects regardless of race, culture, religion, socio-economic status, country of birth or residence (no discrimination); the guarantee of the conditions of justice, respect of equality (in the equal access to health) and of different cultural contexts.

The respect of dignity, physical integrity, autonomy of participants and justice between subjects in accordance to the good clinical practices are ensured through preliminary verification of scientific relevance of research; protection of safety and well-being of participants; equity in the enrolling and selection of participants; balance of reasonable risks compared to potential benefits; expression of informed consent; appropriate treatment during and after the trial; compensation for direct damages to health; distribution of equal burdens and benefits.

The application of general ethical standards of clinical trials to the different cultural context, in particular to developing countries, needs an activity of interpretation and specification. In an ethical framework that recognizes the priority of the human dignity and justice emerges the necessity of additional standards of safeguard to avoid exploitation or abuse of particularly vulnerable populations because of poverty, lack of education and understanding of scientific issues, lack of technical skills, scarce resources, disease, inability to have access to the most basic and essential health products and services.

The process of interpretation might be helped by a community consultation to acquire better knowledge of local culture and involving community representatives in the elaboration of research projects. In this context, the role of the cultural mediator is important. The aim is neither to impose foreign ethical standards nor to adapt to local standards, but to apply generally recognized principles and values taking seriously into account the conditions and needs of the specific culture.

The Universal Declaration on Bioethics and Human Rights underlines some specific additional standards ethically required, explicitly or implicitly.

1   Responsiveness and direct relevance of the clinical trial to the real health needs and specific requirements of the vulnerable population of the host country populations (Article 21 d). International testing should be considered as a priority in relation to the specific interests and priorities of health of the populations of the host country. In this sense, the right to health care as protection of the objective good of a person must be considered a fundamental international right.

2   Enrollment of the subjects should guarantee equity considering the possible advantages of participants in relation to the population and ensuring benefits both to participants and to the population as a whole (Article 21 c and 15). The balancing of risks/benefits should be commensurate with the basic conditions of the population (including nutritional, epidemiological and health conditions), in reference to each individual, but also to the community. Commensuration of risk for the individual and the population in relation to the benefits for "third parties" (with reference to the countries performing the trials) is ethically unacceptable. Research is ethically justified if it provides reasonably direct benefits to participants and indirect benefits for the overall population, with the minimization of risks to people participating in the research, but also for the vulnerable population as a whole.

3   Informed consent should be tailored to local customs, verifying that it is voluntary and freely given without coercion, incentives or 'undue inducement' (Article 5 and 6) (oral and witnessed for the illiterate, with permission of community leader or family involvement). With regard to voluntariness and lack of 'undue' influence, it should be noted that in developing countries participation in a trial could be an advantage for those who have difficulty in obtaining food and basic health care. The socio-economic conditions could push to participate without an adequate awareness of the risks in the research. Another problem could be the difficulty of some populations to grasp the concept of research, which tends to be confused with care and assistance (the so-called therapeutic misconception). The involvement of other persons in the expression of informed consent is acceptable only if there are ways to verify the actual awareness of individual participation (as well as the possibility to withdraw it) and the absence of direct or indirect external pressure. This awareness should be personal and cannot be replaced by others.

4   An issue connected to informed consent is confidentiality (Article 9). Confidentiality may be weakened (if not obliterated) given the family's possible involvement in the process of the permission to research. The fact that in some cultures there is a lack of the concept of "privacy" should also be considered. This raises an ethical problem: the

participation in research may mean, for vulnerable populations, the risk of the stigma of being sick. In this context, cultural associations may play a supportive role, helping the patient not to be marginalized.

5   Appropriate treatment should be guaranteed, ensuring that participants enjoy potential benefits and are compensated for any harm directly related to participation, helping health care infrastructure to support proper distribution and guaranteeing continued access to post-trial benefits and treatment to participants and to the population outside the research context of the country where the trial is conducted, as expression of international cooperation and solidarity (Article 15). This means also that protection should be provided through arrangements of a mandatory insurance in view of possible damages, where the premium is assessed in relation to the local economic state. This could be guaranteed also by independent organizations that are non-profit and internationally accredited, which may have the role of monitoring this ethical requirement.

6   An ethical requirement is the need to assist developing countries in building the capacity to become fuller partners in international research both on scientific and ethical levels, enhancing collaboration and creating an atmosphere of trust and respect. Assistance should be guaranteed to developing countries during the experimentation without inflicting on them the burden of the "indirect costs" of the trial, on an already precarious local health system, and helping them to become full partners in international research, stimulating the improvement of the local health system and transferring technical and scientific skills, involving also doctors and representatives of the host country, to monitor compliance with ethical standards and avoid abuse. It is an ethical requirement of experimentation that the investigators assume responsibility and solidarity in the framework of international cooperation which continues even after the trial, so that research participants do not feel abandoned. In this sense, experimentation is justified to the extent that the product – if it proves effective – can become available to the entire population. There is considerable international debate, even as regards the ways in which this ethical requirement can actually be met.

7   There should also be specific training for doctors and the medical staff conducting this experimentation as well as education involving local doctors and health personnel, often in particularly fragile conditions, so that care becomes a "collaborative partnership" and enables to develop in the host country the skills required to independently conduct clinical trials and ethical assessments (also, possibly, with the institution of local Ethical Committees).

The Universal Declaration on Bioethics and Human Rights constitutes a reference point for the protection of human beings in transnational research, in order to avoid economic interests prevailing over respect of dignity and justice. There is a need for Western countries to realize that advances in scientific knowledge do not mean that we can use them to exploit poor countries for one-way benefit. The intrinsic value of this obligation must be the same for each country and ensure that each country may benefit from the positive results of clinical trials regardless of the level of literacy, wealth, social advancement, techno-scientific progress. This is one of the concrete paths to deliver global justice in health and welfare.

Bibliography

Angell, Marcia 1997, The Ethics of Clinical Research in the Third World, *New England Journal of Medicine*, 337: 847–849.

Angell, Marcia 2006, The Body Hunters, *The New York Review of Books*, 6 Oct. 2005, vol. 52, tr. It. La Rivista dei Libri, marzo, 2006, pp. 40–44.

Baker, Robert 1998, A Theory of International Bioethics: Multiculturalism, Postmodernism, and the Bankruptcy of Fundamentalism, *Kennedy Institute of Ethics Journal*, 8: 201–231.

Benatar, Solomon R. 1998, Imperialism, Research Ethics and Global Health, *Journal of Medical Ethics*, 24: 221–222.

Benatar, Solomon R., Singer, Peter A. 2000, A New Look at International Research Ethics, *British Medical Journal*, 30, 321(7264): 824–826.

Emanuel, Ezekiel J., Wendler, David, Killen, Jack, Grady, Christine 2004, What Makes Clinical Research in Developing Countries Ethical? The Benchmarks of Ethical Research, *The Journal of Infectious Diseases*, 189(5): 930-937.

European Commission, *The Role of Ethics in International Biomedical Research* 2010, Report of the 2nd meeting of the European's Commission's International Dialogue on Bioethics, Madrid 4–5 March 2010, Bureau of European Policy Advisor, Publications Office of the European Union, Luxembourg.

Falk, Richard 1999, *Predatory Globalization: A Critique*, New York: Polity Press.

Glickman, Seth W., McHutchison, John G., Peterson, Eric D., Cairns, Charles B., Harrington, Robert A., Califf, Robert M. 2009, Ethical and Scientific Implications of the Globalization of Clinical Research, *New England Journal of Medicine*, 360: 816–823.

Hawkins, John S., Emanuel, Ezekiel J. (eds.) 2008, *Exploitation and Developing Countries: The Ethics of Clinical Research*, Princeton, NJ: Princeton University Press.

Hyder, Adnan A., Wali, Salman A., Khan, Agha N., Teoh, Noreen B., Kass, Nancy E., Dawson, Liza 2004, Ethical Review of Health Research: A Perspective from Developing Country Researchers, *Journal of Medical Ethics*, 30: 68–72.

Koski, Greg, Nightingale, Stuart L. 2001, Research Involving Human Subjects in Developing Countries, *New England Journal of Medicine*, 345(2): 136–138.

Lorenzo, Cláudio, Garrafa, Volnei, Solbakk, Jan Helge, Vidal, Susana 2010, Hidden Risks Associated with Clinical Trials in Developing Countries, *Journal of Medical Ethics*, 36: 111–115.

Macklin, Ruth 2004, *Double Standards in Medical Research in Developing Countries*, Cambridge: Cambridge University Press.

Pavone, Ilja Richard 2016, Legal responses to placebo-controlled trials in developing countries, *Global Bioethics*, 27: 76–90.

Petryna, Adriana 2005, Ethical Variability: Drug Development and Globalizing Clinical Trials, *American Ethnologist*, 32(2): 183–197.

Shapiro, Harold T., Meslin, Eric M. 2001, Ethical Issues in the Design and Conduct of Clinical Trials in Developing Countries, *New England Journal of Medicine*, 345(2): 139–142.

Varmus, Harold, Satcher, David 1997, Ethical Complexities of Conducting Research in Developing Countries, *New England Journal of Medicine*, 337: 1003–1005.

Wendler, David, Emanuel, Ezekiel J., Lie, Reidar K. 2004, The Standard of Care Debate: Can Research in Developing Countries be Both Ethical and Responsive to Those Countries' Health Needs?, *America Journal of Public Health*, 94: 923–928.

International norms

Council of Europe, Convention for the Protection of Human Rights and Dignity of the Human Being with Regard to the Application of Biology and Medicine: Convention on Human Rights and Biomedicine of the Steering Committee on Bioethics of the Council of Europe (1997): human dignity (Article 1), and the primacy of human well-being over the sole interest of science and society (Article 2), equity of access to healthcare (Article 3), free and informed consent (Article 5), the protection of the people that lend themselves to research (Articles 16–17).

European Union, The Charter of Fundamental Rights (2000): human dignity (Article 1), the right to personal integrity, the respect of free consent, the prohibition of exploitation of the body (Article 3).

Council of Europe, The Additional Protocol Concerning Biomedical Research of the Convention on Human Rights and Biomedicine (2005): Article 29 refers to the multi-center research and the duty to apply one standard of ethical evaluation.

Regulation (EU) No 536/2014 of the European Parliament and the Council of 16 April 2014 on clinical trials on medicinal products for human use, and repealing Directive 2001/20/EC.

International guidelines

International Conference on Harmonization on Technical Requirements for Registration of Pharmaceuticals for Human Use (ICH), *Guideline for Good Clinical Practice* (ICH Harmonised Tripartite Guideline) (1997).

Council for International Organizations of Medical Sciences (CIOMS) in collaboration with the World Health Organization (WHO, *International Ethical Guidelines for Biomedical Research Involving Human Subjects* (2002).

World Medical Association, *Declaration of Helsinki: Ethical Principles for Medical Research Involving Human Subjects* (adopted in 1964, revised in 1975, 1983, 1989, 1996, 2000, 2008, 2013).

Reports and opinions of national bodies

Comité Consultatif National d'Etique pour les Sciences de la Vie et de la Santé, La coopération dans le domaine de la recherche et équipes françaises biomedical entre équipes de pays en voie de développement économique. Rapport (1993).

National Bioethics Advisory Commission, Ethical and Policy Issues in International Research: Clinical Trials in Developing Countries, Report and Recommendations, Bethesda, Maryland, Vol. I, 2001.

Nuffield Council on Bioethics, The Ethics of Research Related to Healthcare in Developing Countries, 2002.

European Group on Ethics in Science and New Technologies, Ethical Aspects of Clinical Research in Developing Countries, 2003.

Food and Drug Administration, Report of the Department of Health and Human Services, Human Subject Protection; Foreign Clinical Studies not Conducted under an Investigational New Drug Application, Federal Register, Vol. 73, No. 82, 2008.

Comitato Nazionale per la Bioetica italiano, Pharmacological trials in developing countries 2011.

VII In search for bioethical international standards

The legal perspective of the Universal Declaration on Bioethics and Human Rights

Carmela Decaro Bonella and Francesco Alicino

Introduction

One of the most recent features of the emerging international tools relating to bioethics is that it assigns an important role to human dignity and human rights. UNESCO's Universal Declaration on Bioethics and Human Rights (hereafter the Declaration) is the best example of that. It is in fact a result and a manifestation of a deeper concern, closely connected to the mission of UNESCO: the recognition and the protection of human dignity and human rights are seen as guarantees that progress in science will contribute to peace, security, and prosperity. This, on the other hand, underscores the problems arising from the pressing process of globalization and, consequently, what is called global bioethics, which also considers structural injustices and social inequalities, particularly in health, health care, and environment (Solinís 2015).

The globalization of health care, medical research, and the technology has created a different context to such an extent that the major bioethical concern is no longer only the power of science and technology. It is also the global health, discrimination, and the negative effects of higher levels of income disparities, contemplated as fundamental problems of our era. These are in effect problems that, in the light of the protection of human dignity and human rights, produce contradictory and unsustainable situations, such as those referring to the persistence of readily treatable diseases and the development of science and economic inequality on a world scale (Farmer 2003: 192).

In this sense, despite the great number of existing international documents relating to bioethical issues, the 2005 Declaration is the first instrument that comprehensively deals with the linkage between human rights, human dignity, and bioethics (Macpherson 2007: 588–590). In addition, with the Declaration virtually all of UNESCO's Member States have reached an agreement in a very sensitive area.

The 2005 document seems nonetheless to cause difficulties in relation to the interpretation and implementation of its provisions. The Declaration is in fact a typical 'soft law' agreement, which does not have a binding effect *per se*. It is conceived to have such effect in the long term, during which Member States play a very crucial role, as explained by Andorno in Chapter I of this volume. Its compliance mechanism is in other words limited to States' self-reporting of their implementing steps. Besides, the connection between human rights, human dignity, and bioethics appears to be consistent with the Declarations' main objective: the development of international normative standards in the area of bioethics.[1] Sometimes,

---

1 Following a feasibility study to assess the possibility of elaborating an international instrument on bioethics, the 32nd UNESCO General Conference considered it opportune and desirable for UNESCO "to

though, it fails in practice at achieving this goal because of principles that are too broad and appear to lack of *genus proximum et differentia specifica*.[2]

It is true that the Declaration includes powerful principles, such as human rights and human dignity. It is also true, however, that these principles denote slippery ideas, particularly when relating to the international arena and global bioethics (Gert 2014: 13–27). That is the case of the 'universality' of human rights and human dignity, largely affirmed in the Declaration, but not clearly explicated.

After some remarks about the sense of the existing global bioethics, in this Chapter we will focus attention on the form and the content of the Declaration (this issue will be further discussed in Chapter VIII). This is in fact a result of a compromise, which implies an interaction between 'bioethics' and 'biopolicy'.[3] In this perspective, we will analyse the practical effects of the 2005 document. In particular, we will try to understand if the linkage between human rights, human dignity, and bioethics is beneficial for affirming the so-called global bioethical standards that, at the end of the day, is the primary objective of the 2005 Declaration.

The need for a global bioethics

Today's growing appeal of the bioethics may be attributed to the empowering combination of two important notions, traditionally linked to moral philosophy and philosophy of law: 'application' and 'principle' (Stanton-Jean 2016: 14). In this field application has a double connotation. It indicates that bioethics is available for what we usually do. As a consequence, it tries to affirm practical implications, in order to resolve our daily problems. In this sense, bioethics can contribute to:

- the clarification of everyday difficulties arising not only in health and health care, but also in other fields, such as those referring to socio-economic inequalities and environmental issues;
- the analysis of ethical dilemmas;
- the resolution of complex cases.

The second feature is focused on 'principles'. If bioethics is conceived as applied ethics in some sensitive areas, then subsequent reflection is needed on what is being applied. In this case the first thing to say is that bioethics should include principles coherent with the moralities of obligation or the 'moral objects' in life sciences, medicine, and technology, which have dominated modern ethical discourse (Stout 2004: 163). From a juridical point of view, these 'objects' reflect into the idea of contemporary constitutionalism implying the recognition and the protection of human dignity and human rights that, as many know, are strictly related to the recent history of the Western legal system (Alicino 2010: 1–32).

It should be noted that, with the rapid dissemination of information and the pressing process of globalization, the bioethical questions – and the relative moral objects – are now more frequently and openly discussed. They are no longer the exclusive concern of scientists and medical professionals. They involve all the people and society as a whole, including

set universal standards in the field of bioethics for due regards for human dignity and human rights and freedoms, in the spirit of cultural pluralism inherent in bioethics" (UNESCO, 2004: 45–46).

2 See *infra*, paras. 3–4.

3 For these notions see *infra*, para. 4.

future generations. Disease, disability, death, suffering, and environment are human experiences that sooner or later affect everybody. For these same reasons, bioethics is no longer just an academic field, as policy makers too are drawn into examining the questions related to medicine, life sciences, and technology advances.

On the other hand, because of globalization, these advances spread around the globe, bringing with them new bioethical queries. Take for example the cloning techniques that, even if developed in some States, can be applied in others, irrespective of the national law applicable. At the same time, and from a different perspective, diverse bioethical issues arise because of inequality and injustice: if, for instance, an effective medication for diseases (such as HIV, malaria, and tuberculosis) is available in some countries, it is ethically problematic when patients die in others because of a lack of economic resources (Pavone 2012: 65).

This shows that technological progress, new knowledge, new diagnostics, preventive and therapeutic interventions and their applications have significantly changed medicine, life sciences, and health care, giving rise to new ethical dilemmas in highly developed and less developed States. In addition, the existence of global markets has created new problems such as organ trade, medical tourism, corruption, bioterrorism, pandemics, malnutrition, hunger, and climate change. And, even though such troubles exist only in some local-regional contexts, many times they also have serious consequences for the rest of the world.

It is a fact that, while some bioethical problems have expanded to operate on an international and global scale, rules governing such problems are still based on the national perspective. Nevertheless, in the last two decades new bioethical queries have demonstrated that national legislations or regulations are no longer sufficient – if not ineffective – in this field. The global character of contemporary science and technology, increasing number of researches coming from different nations, social-economic inequalities, environmental damages, and new transnational crimes have indeed produced the need for a worldwide approach in bioethics.

Traditional bioethical issues are now confronted with new challenges. Hence, it is true that bioethics may have primarily originated in Western States, but it is also true that its implications are relevant on a global scale. For this, bioethical discourse can no longer be referred only to national territories. It has *de facto* become supranational, involving the concerns of all human beings wherever they are. This means that, in the perspective of contemporary bioethics, it is not sufficient to export its principles to non-Western countries. Rather, in order to address the new transnational tasks, bioethics needs to rethink its West-centered approach without, however, losing sight of the role of the States' laws, except that they need now to be harmonized. The new context implies supranational action designated to affirm shared principles and provisions, supported by a practical cooperation for their interpretation and application.

From here stems the contribution of UNESCO that, especially in the last two decades, has tried to bring about a real change in this field. In doing so, it has generally approved recommendations and declarations, thus proposing to Member States principles that are susceptible of inspiring national legislations, guidelines, and/or regulations. Its main goal is to provide a 'universal' (or at least a more globalized) understanding of bioethical issues. The 2005 Declaration could be considered as one of the most important steps into this direction.

The legal form

The Declaration constitutes an original initiative, which combines bioethics, human dignity, and human rights, thus sustaining the development of a supranational approach in this field (Boussard 2009: 293). The Declaration, though, 'invites' Member States to promote, rather than to implement, its content.[4] In positivist international law, only binding documents are to be implemented. The 2005 document is undoubtedly a non-binding text. It falls in the ambit of quasi-legal instruments that, promoting some national policies, is in line with the State-centered approach of 'international law', the body of rules that traditionally governs the relations between or among nations.

Member States, which have not already done so, are encouraged to establish independent, multidisciplinary, and pluralist ethics committees,[5] to promote informed pluralistic public debate, to foster bioethics education and training, and to take appropriate legal measures to facilitate transnational researches.[6] International organizations such as UNESCO will continue to assist countries to develop an ethical infrastructure, so that human beings everywhere can benefit from the advances of science and technology within a framework of respect for human dignity, human rights, and fundamental freedoms.[7] Some provisions of the Declaration aim in any case at giving its principles the greatest audience possible, also highlighting the role of non-State actors. In this sense, it is significant that Member States adopting the Declaration have committed themselves to encourage the establishment of national ethics committees as a way of fostering information and knowledge in bioethics.

The Declaration is in fact addressed to States, giving them primacy in the implementation of its provisions; what Articles 22–24 precisely establish. On the other hand, however, the 2005 document "provides guidance to decisions or practices of individuals, groups, communities, institutions and corporations, public and private";[8] what carries the original intent of the drafters, scientific experts who are members of the International Bioethics Committee (IBC) (Boussard 2009: 294–295). It should be noted that between these two approaches there is not discrepancy, as it might appear to be at first sight. The 2005 document is indeed in line with precedent international human rights law, whose provisions are seen as a common standard of achievement for all people and all nations: they seek to enlist every individual and every organ of society in a universal attitude.

In sum, from a legal point of view, it is clear that the 2005 Declaration is not a binding instrument. Nonetheless, the recent history of universal human rights demonstrates that what starts as soft law can in time obtain a more effective impact; as is the case, for instance, with regard to the 1948 Universal Declaration on Human Rights. This means that the 2005 Declaration may in future become an incentive for other national and supranational initiatives, leading to the drafting of legally binding documents, such as the European Convention on Human Rights and Biomedicine, adopted in the regional context of the Council of Europe (Council of Europe 1997). Otherwise, the 2005 Declaration could at least become a point of reference for national and regional jurisprudence: it is not by chance that it has been cited as a relevant text in some important case-laws, like those referring to the

---

4 See Article 22, where it is affirmed that "States should take all appropriate measures, whether of a legislative, administrative or other character, to give effect to the principles set out in this Declaration in accordance with international human rights law. Such measures should be supported by action in the spheres of education, training and public information".

5 Article 23.2.

6 Article 24.

7 See Article 25: "1. UNESCO shall promote and disseminate the principles set out in this Declaration. In doing so, UNESCO should seek the help and assistance of the Intergovernmental Bioethics Committee (IGBC) and the International Bioethics Committee (IBC). 2. UNESCO shall reaffirm its commitment to dealing with bioethics and to promoting collaboration between IGBC and IBC".

8 Article 1.

European Court of Human Rights (European Court of Human Rights 2006: para. 42) (Pavone 2009: 101).

However, all of this says that the 2005 document is the beginning – rather than the end – of the process of standard-setting of bioethics on the global (universal) scale. So, in order to better evaluate the real impact of the Declaration, special attention needs to be given to the application and the practical implications of its provisions at local, national, and regional levels.

The substantive content

From a general point of view, the Declaration establishes principles determining the different obligations and responsibilities of the 'moral agent' in relation to different categories of persons, their dignity and their environment. As a consequence, these principles are arranged according to a gradual widening of the range of 'fundamental needs': human dignity, human rights, sharing of benefits, limiting harms, autonomy, consent, privacy, equality, respect for cultural diversity, solidarity, social responsibility, protection of the biosphere and biodiversity. In the light of these needs, the Declaration tries to balance individualist and communitarian perspectives, which are evaluated from at least three sectors: medicine and health care; social-economic background; and environment. In other words, the Declaration addresses "ethical issues related to medicine, life sciences and associated technologies as applied to human beings, taking into account their social, legal and environmental dimensions".[9] From here stem the multiple aims of the 2005 document, which includes the necessity to provide "a universal framework of principles and procedures to guide States in the formulation of their legislation, policies or other instruments in the field of bioethics".[10]

More specifically, the Declaration recognizes the principle of autonomy of persons to make decision,[11] as well as the principle of solidarity among human beings and international cooperation.[12] It emphasizes the principle of social responsibility and health,[13] which aims at orienting bioethical decision-making towards urgent issues, such as access to quality health care, essential medicines for weaker subjects (poor people, children, and women, for example) adequate nutrition and water, reduction of poverty and illiteracy, and improvement of living conditions. In order to advance decision-making, the Declaration's principles are to be understood as complementary and interrelated: each norm shall be considered in the context of the others, as appropriate and relevant in the circumstances.[14] This said, the most important thing is that the Declaration anchors its bioethical rules in principles referring to human dignity, which also implies the recognition and protection of human rights and fundamental freedoms.[15]

The section on the 'application' (Articles 18–21) is also innovative because it provides the spirit in which the Declaration's provisions ought to be applied. It calls for professionalism, honesty, integrity, and transparency in the decision-making process, appropriate assessment-management of risk, and ethical transnational practices that help in avoiding

9 Article 1.
10 Article 2.
11 Article 5.
12 Article 13.
13 Article 14.
14 Article 26.
15 See *infra*.

exploitation of countries that do not have an ethical infrastructure. If the application of the Declaration's principles is to be limited at the national level, only legislation or statutory law should do it. This applies as well for very sensitive areas, such as public safety, detection and prosecution of criminal offences, protection of public health, and protection of rights of others. In any case, such legislation needs to be consistent with international human rights law,[16] which means that "[n]othing in this Declaration may be interpreted as implying for any State, group or person any claim to engage in any activity or to perform any act contrary to human rights, fundamental freedoms and human dignity".[17]

A compromise between bioethics and biopolicy

As can be easily noticed, the form and content of the Declaration is necessarily a result of compromise, which is a normal way of proceeding during supranational negotiations. In this particular case, however, the compromise is affirmed not only between nations, with their different socio-cultural, religious, and legal traditions. It also involves the dialectic relationship between science and politics or, to be more accurate, between bioethics (based on the relevance of evidence in the life of science and technology) and biopolicy (based on the behaviour of policy makers in sensitive matters) (Somit and Peterson 2012: 3–12). This, on the other hand, shows that bioethical issues are no longer the exclusive concern of scientists and medical professionals (Plomer 2005: 23–42). They also entangle fundamental aspects of everyday life and, as such, imply some of the most important tasks of policy makers. All of this is even more evident by the fact that the IBC's experts drafted the preliminary text of the Declaration, but the governmental authorities made the ultimate decision on it (Magnus 2016: 30). In this manner, the linkage between science and politics reflected into the content and form of the 2005 document.

The product of this double interaction can certainly be seen as a vital characteristic of the Declaration, which enjoys the support of virtually all UNESCO's Member States with a range of diverse ethical views: UNESCO has in effect been able to manage very different positions, find common grounds among various traditions, and indicate minimum standards for bioethics that could be universally acceptable (Andorno 2009: 223–240). Nevertheless, the double interaction might also reflect into the Declaration's weakness, which is mainly due to the vagueness, if not the ambiguity, of its principles. These two conflicting interpretations (vital characteristic *vs* weakness) can in any case be explained by the fact that political logic does not always coincide with scientific rationality, as, from the opposite perspective, the structure and argumentation provided by scientific experts are not always completely implemented by political negotiations. After all, it is not an accident that these two diverging interpretations are particularly evident in the light of the 'aims' of the Declaration, especially when related to the 'general aims' of bioethics (Levitt and Zwart 2009: 367–377; Schuklenk 2010: 83–87).

The aforementioned interaction has in fact led UNESCO to the adoption of indeterminate but, and at the same time, very powerful concepts, like those referring to human rights and human dignity. Take, for example, Article 2(c), where it is stated that one of the most important aims of the Declaration is "to promote respect for human dignity and protect human rights, by ensuring respect for the life of human beings, and fundamental freedoms, consistent with international human rights law". So, in this sense, the 2005 document

16 Article 27.
17 Article 28.

recognizes the importance of freedom of scientific research and the benefits derived from technological developments. But it also stresses "the need for such research and developments to occur within the framework of ethical principles set out in this Declaration and to respect human dignity, human rights and fundamental freedoms".[18] Even Article 3, specifically devoted to human dignity and human rights, simply affirms that they (human dignity and human rights) "are to be fully respected". And that is all. Not to mention Article 12 which, underling the "importance of cultural diversity and pluralism", affirms that these considerations are in any case "not to be invoked to infringe upon human dignity, human rights and fundamental freedoms".

In sum, the 2005 document puts human dignity and human rights at the centre of its principles without, however, explicating their normative content and concrete impact. It seems that we should accept them as fundamental axioms of the Declaration and its bioethical system. One can always say that we might take human dignity and human rights as universally used. This solution, though, is not without problems, notably when related to the inter-supranational arena. In this context some policy makers use the rhetoric of human dignity to argue, for instance, against gay rights, access to medically assisted procreation, voluntary euthanasia, and a number of additional purposes. Conversely, others use the same human dignity–related rhetoric, but they come to diametrically opposed conclusions.

To give a more specific example, in situations like abortion we should wonder whether the life of the parent bears the same dignity as that of the newborn child. Based on the viewpoint that all forms of life relating to human beings are sacred, in some legal traditions the foetus must enjoys the same rights as the mother. And this explains why, from this ethical perspective, abortion is forbidden: here the foetus is considered a human being from the very beginning of the pregnancy or at least from an early stage of the pregnancy. Another position, though, will say that parental dignity is not equivalent to that of the newborn. And, having said that, it is easier to grant the mother the option of undergoing an abortion. Now, on the basis of the 2005 Declaration, some States pass laws on this issue, sometimes creating a balance between these different viewpoints. However, for others the 2005 document argues that unborn infants have no right for dignity (Schmidt 2007: 578–584).

Article 6(1) of the Declaration gives us another important example. It states that part of a patient's dignity is expressed in the doctors' obligation to obtain informed consent to the proposed medical procedure, with the exception of life-threatening emergencies (where it is impossible to obtain the patient's informed consent):

> [a]ny preventive, diagnostic and therapeutic medical intervention is only to be carried out with the prior, free and informed consent of the person concerned, based on adequate information. The consent should, where appropriate, be express and may be withdrawn by the person concerned at any time and for any reason without disadvantage or prejudice.

Because consent is considered an expression of human dignity, a broad-ranging informed consent is not ethical. Nevertheless, there can be situations where obtaining consent can harm the patient and detract from his or her dignity as well. Every situation and every patient should therefore be regarded individually and according to the particular circumstances of the case. It means that human dignity cannot be treated as a constant and immutable value.

---

18 Article 2(d).

Which is also proved by the fact that there are societies where it is acceptable for close relatives (parents, spouses) to take part in the decision-making process and where talking to them about the patient is part of the culture (UNESCO 2011: 86).

This leads to another remark. Although human rights were conceived with the individual in mind, there is a strong trend towards collective rights, such as the right to speak one's native language and educate children in that language, the right to cultural preservation, the right to national self-determination, the right to peace, the right to a healthy environment, and so on. Now, in this perspective UNESCO understands that, for example, individuals have the right to refuse vaccinations. But it also affirms that society has the right to vaccinate its citizens in order to promote health. On this point, Article 3(2) of the Declaration declares that "[t]he interests and welfare of the individual should have priority over the sole interest of science or society".[19] At the same time, though, Article 14(1) states that "[t]he promotion of health and social development for their people is a central purpose of governments that all sectors of society share". Vaccination of the population reduces rates of illness and therefore it is expected that public interest dictate policies supporting vaccination. Considered the most effective treatment for some diseases today, vaccination is in any case not an obligation: people who do not desire vaccination will usually not receive it; they must remember, though, that there may be consequences to non-treatment. Yet there could be circumstances in which individual interests do not take priority over public interests. So, preserving public welfare and reducing diseases are important values, and when they conflict with the individual's right to refuse treatment, a balance must be struck between these two needs. As a consequence, each situation should be evaluated on its own merits and each society, State, and government must define a balanced policy based on its values (UNESCO 2011: 114–115).

These examples make it evident that, when linked with bioethical issues, human dignity and human rights are not perhaps so universal as they might seem. Moreover, from this point of view the 2005 Declaration mirrors the phenomenon of agreeing on the lowest common denominator, which implies principles that everybody agrees upon, ignoring items about which there is no an 'universal' consensus.[20] The other way to reach an agreement in matters like these is to choose formulations that are sufficiently vague and that each ethical system can accept; what might include that each system can interpret them consistently with its own cultural-religious vision. UNESCO surely is not arguing that everyone can use the Declaration in random interpretation. Given the vague definition of its principles, this is nevertheless a scenario that needs to be taken into serious consideration.

Additionally, UNESCO accepts that academic bioethicists are probably not too impressed with the scientific substance of the Declaration, which is in fact addressed to Member States and their policy makers. Yet this might precisely be another problem. We should not forget that the 2005 document targets bioethical issues and that, as such, includes uninformed statements on research ethics. Hence, harms would be caused if some Member States were to make the Declaration's principles the foundation of their own laws, but on the basis of an 'unscrupulous' – that is to say 'unscientific' – interpretation. Perhaps, in cases like these it is not certain that UNESCO may oppose it effectively.[21] This is because of the vagueness of

---

19 On the problems about the interpretation of this Article see Schuklenk (2010: 84–85).
20 The Declaration was in effect adopted unanimously, without any contrary votes or recorded abstentions (Kirby 2008–2009: 309–331).
21 Indeed, as an intergovernmental third party, UNESCO is in a unique position to ensure that while the international progress of science and technology are not unduly impeded, research participants worldwide are also properly protected.

the Declaration's principles and, consequently, the different interpretative possibilities that can be extracted from them: selecting among these possibilities, one can always say that my interpretation is as licit and appropriate as yours.

In brief, the Declaration was formulated with the aim of achieving consensus from all UNESCO Member States. Although initially drafted by the bioethics experts in the IBC, the contents were subject to substantial editing by governmental officials, whose negotiation produced an instrument with principles framed at a very high level of abstraction. The Declaration has thus been criticized for being "at best a toothless statement of vague principles, and at worst a potential source of mischief that will harm research and public health efforts" (Wolinsky 2006: 355).

The reasons for adopting human dignity and human rights

A this point we have to understand the reasons behind the striking insistence on human rights and human dignity that can be found in the Declaration. Surprisingly, and despite all the aforementioned problems, the answer is quite simple: biomedical effects are closely related to basic prerogatives of human beings, individually or collectively. Hence, if human dignity and human rights are generally recognized as the foundation of these prerogatives, then it is normal that they are mentioned as the ultimate rationale of legal frameworks for regulating practices in biomedical technology.

In this manner, the notions of dignity and human rights can at least be considered as the last barrier against the alteration of some basic characteristics of the human beings, which might result from some disgraceful practices; like those referring to the painful and often deadly experiments on thousands prisoners conducted by a number of Nazi doctors in the first half of the last century. It is not by chance that the 1997 Universal Declaration on Human Genome and Human Rights directly appeals to the notion of human dignity in order to reject such practices.[22] Moreover, in the light of global perspective, human dignity might be considered as one of the most important principles of bioethics.

While there is no clear definition for it, dignity reflects into the need to promote respect for the intrinsic value of every human being. This is coherent with the role played by the concept in both the 1948 Universal Declaration of Human Rights (UDHR) and the Charter of Fundamental Rights of the European Union (the EU Charter) where, among other things, it is stated that "[a]ll human beings are born free and equal in dignity and rights"[23] and that "[h]uman dignity is inviolable": as such, "[i]t must be respected and protected".[24] Thus, human dignity is here considered an overarching principle, which is normally accompanied by other effective and practical human rights (Kirby 2008–2009: 309–331).

The dignity of the human person is not only a fundamental right in itself. It is also a foundation for subsequent freedoms and rights. Take for example the right to privacy, as

22 See Articles 11 ("practices which are contrary to human dignity, such as reproductive cloning of human beings, shall not be permitted. States and competent international organizations are invited to co-operate in identifying such practices and in taking, at national or international level, the measures necessary to ensure that the principles set out in this Declaration are respected") and 24 ("the International Bioethics Committee of UNESCO should make recommendations, in accordance with UNESCO's statutory procedures, addressed to the General Conference and give advice concerning the follow-up of this Declaration, in particular regarding the identification of practices that could be contrary to human dignity, such as germ-line interventions") of the Universal Declaration on Human Genome and Human Rights.
23 Article 1 of UDHR.
24 Article 1 The EU Charter.

stated in Article 88.2 of the General Data Protection Regulation, approved by the European Parliament on 14 April 2016.

> Member States may, by law or by collective agreements, provide for more specific rules to ensure the protection of the rights and freedoms in respect of the processing of employees' personal data in the employment context. . . . Those rules shall include suitable and specific measures to safeguard the data subject's *human dignity*, legitimate interests and fundamental rights, with particular regard to the transparency of processing, the transfer of personal data within a group of undertakings, or a group of enterprises engaged in a joint economic activity and monitoring systems at the work place.
> (Council of the European Union 2016; my italics)

This implies that right to privacy actually enables individuals to maintain their autonomy and live as they want. More specifically, as stated in Article 9 of the 2005 Declaration,

> [t]he privacy of the persons concerned and the confidentiality of their personal information should be respected. To the greatest extent possible, such information should not be used or disclosed for purposes other than those for which it was collected or consented to, consistent with international law, in particular international human rights law.

In this sense, the right to privacy is an integral part of human dignity (European Data Protection Supervisor 2015), which means that the protection of that right "should be based directly on the protection of human dignity" (Floridi 2016: 2). The right to privacy, though, sometime conflicts with other rights, such as the right of other persons to know the truth about their health. In such situations, we must then find a reasonable balance between different prerogatives. And we should do this on the basis of the protection of human dignity.

Although UNESCO's rules concerning bioethics recognize a central role to human dignity, they also underscore the fact that this principle alone cannot solve most bioethical dilemmas. Human dignity is not a magic formula that can be invoked to find a precise solution to the complex challenges posed by current bioethical issues. In order to become functional, it needs other – and perhaps more – concrete concepts, like those referring to privacy, informed consent, physical integrity, equality, solidarity, non-discrimination. Not for nothing, these concepts are normally formulated by using the terminology of human rights. Here the example is given by Article 3.1 of the 2005 Declaration, which adds "human rights and fundamental liberties" when referring to human dignity. This indicates that one of the most important achievements of the Declaration consists precisely in having "integrated the bioethical analysis into a human rights framework" (Kirby 2005: 126).

Certainly, it is not easy to define human dignity in clear and unambiguous terms. But the same happens with other basic conceptions, such as justice and freedom. However, it would seem unreasonable to argue that we should abandon these important notions. It is true that the bioethical debates show often an inflationary use of the expression 'human dignity'. And this would invite us to avoid it, especially when no additional explanation is given to make it clear why a particular practice is regarded as being in conformity – or not – with dignity (Schüklenk 2010: 83–87). Nonetheless, the abusive rhetoric surrounding this principle also reflects concerns about the need to ensure respect for the "human exceptionalism" (Floridi 2016: 5). And this remains a solid argument in favour of the protection of human rights via the principle of human dignity that, on the other hand, is by far broader than simply ensuring respect for 'autonomy'. Indeed, the human dignity–related principle also includes the protection of those who are not yet or no longer autonomous, such as newborn babies and persons suffering from serious mental disorders.

Besides, the connection between human rights, human dignity, and bioethical issues lets us go beyond individual concerns and focus the attention on some practices that risk harming humankind as a whole, including future generations. This is the reason why the 2005 Declaration appeals to the notions of human rights and human dignity to identify "emerging challenges in science and technology taking into account the responsibility of the present generations towards future generations".[25] But this is also the reason why the Declaration has broadened the scope of bioethics to include considerations about the environment, biosphere, and biodiversity: it is no longer possible to advance science and technology without reflecting on the impact of our actions on the environment and other living beings.[26] Hence, unlike traditional bioethics discourse, which tends to place the emphasis mainly on individual persons, the reference to human dignity and human rights leads the Declaration to underline the new context of the global bioethics, where all sectors of society can play an important part in ensuring the ethical conduct of biomedical research and clinical practice. This explains the principles of solidarity and internal cooperation,[27] which are further supported by other principles, like those related to social responsibility and sharing of benefits.[28]

For these reasons, and despite the difficulties offered by its notion, the use of a human dignity framework might facilitate the formulation of supranational (and it is hoped universal) standards in the field of bioethics. In such a sensitive field, where socio-cultural and religious-legal traditions come into play, the feature of human dignity should in any case not be underestimated.

A step into the process of global bioethics

After all, the reference to human dignity and human rights is not an entirely new approach in bioethics (Andorno 2009: 223). For example, the 2003/69 Human Rights and Bioethics Resolution of the UN Commission on Human Rights repeatedly mentions the "dignity of the human being", also recalling that,

> according to the Universal Declaration of Human Rights, the International Covenants on Human Rights and other human rights instruments, recognition of the inherent dignity and of the equal and inalienable rights of all members of the human family is the foundation of freedom, justice and peace in the world.
> (Office of the High Commissioner for Human Rights 2003)

---

25 Preamble of the 2005 Declaration. See also Articles 2 ("[t]he aims of this Declaration are: . . . (g) to safeguard and promote the interests of the present and future generations") and 16 ("[t]he impact of life sciences on future generations, including on their genetic constitution, should be given due regard") of the Declaration.

26 See Article 17 of the 2005 Declaration: "[d]ue regard is to be given to the interconnection between human beings and other forms of life, to the importance of appropriate access and utilization of biological and genetic resources, to respect for traditional knowledge and to the role of human beings in the protection of the environment, the biosphere and biodiversity".

27 See Articles 13 and 24 of the 2005 Declaration.

28 See Articles 14 and 15 of the 2005 Declaration.

Likewise, the Declaration of Helsinki on Research Involving Human Subjects[29] states that "the duty of physicians who are involved in medical research [is] to protect the life, health, dignity, integrity, right to self-determination, privacy, and confidentiality of personal information of research subjects" (World Medical Association 1964). Moreover, the 1996 Statement on the Principled Conduct of Genetics Research – adopted by the Human Genome Organisation; Ethical, Legal, and Social Issues Committee – adheres to international norms of human rights, accepting and upholding "human dignity and freedom" (HUGO 1996). Even Article 1 of the Convention on Human Rights and Biomedicine – approved by the Council of Europe on 14 April 1997 – affirms that

> Parties to this Convention shall protect the dignity and identity of all human beings and guarantee everyone, without discrimination, respect for their integrity and other rights and fundamental freedoms with regard to the application of biology and medicine.
>
> (Council of Europe 1997)

In this sense, the 2005 Declaration can be seen as a fundamental step into a wider process of internationalization of the bioethical rules, which puts at the centre of its actions the protection of human dignity that, in turn, is at the core of human rights. For a more accurate application of the espoused principles, the Declaration should be then read in the light of this process and together with supplementary advices; such as reports elaborated by the IBC, including those on Consent (IBC 2008), Social Responsibility and Health (IBC 2010), Respect for Human Vulnerability and Personal Integrity (IBC 2013) as well as the 2014 report concerning non-discrimination and non-stigmatization. By expounding on the principles, these documents offer a more comprehensive understanding of the 2005 Declaration.[30]

It should be also noted that Article 19 of the Declaration calls for the establishment of ethics committees at various levels. In particular, it affirms that independent, multidisciplinary and pluralist ethics committees should be established, promoted, and supported in order to:

> (a) assess the relevant ethical, legal, scientific and social issues related to research projects involving human beings; (b) provide advice on ethical problems in clinical settings; (c) assess scientific and technological developments, formulate recommendations and contribute to the preparation of guidelines on issues within the scope of this Declaration; (d) foster debate, education and public awareness of, and engagement in, bioethics.

As matter of fact, the need for such bodies flows from increasingly rapid scientific progress and the new possibilities that, on the other hand, create tensions between what can and

---

29 This Declaration was adopted by the 18th World Medical Association (WMA) General Assembly on June 1964. It has been amended many times. The last amendment was approved by the 64th WMA General Assembly, Fortaleza, Brazil, October 2013.

30 In 2009 UNESCO also published a book entitled *The UNESCO Universal Declaration on Bioethics and Human Rights: Background, Principles and Applications* (UNESCO 2009). This book provides a thorough examination article-by-article of the Declaration, explaining how it could be used as a tool to address ethical issues. As almost all of the authors were involved in the elaboration of the Declaration, their contributions reveal its historical background and the potential interpretation and application of its principles, including those referring to human dignity and human rights.

what may be done in regard to the protection of human dignity and human rights. The traditional principles of medical ethics are not providing the necessary answers. Thus, the core mission of the national bioethics committees (NBCs) is to examine these issues, in an interdisciplinary effort, ascertaining what constitutes responsible action at the interface of the biological sciences, medicine, and health care. NBCs endeavour to clarify issues and produce ethical judgements that are conducive to discussion. Their opinions are meant to foster debate and ultimately contribute to the well-being of the people concerned. They do not provide ready-made answers: NBCs' goal is not to lay down the supposedly only politically or morally correct positions for Member States. They must instead make a substantial contribution to the discussion among the public and the authorities (UNESCO 2010). In this manner, NBCs play an important role in establishing global bioethics norms and rules, by stimulating awareness for all stakeholders (IBC 2013b).

Hence, from bottom to top and from top to bottom, the aforementioned process of internationalization of the bioethical rules can contribute to a greater respect for human dignity and human rights, by facilitating the participation of citizens in decisions that directly or indirectly affect them. For these reasons, since the adoption of the Declaration, assistance to some Member States in establishing NBCs and training their components has been a major part of the work of UNESCO in this very sensitive and strategic field.

Conclusion

Born in the 1970s, bioethics was traditionally conceived as a response to the power of medical science and technology. It expanded as a public discourse empowering individual citizens and encouraging States' law in some areas as research, transplantation, reproduction, and end-of-life care. Since then, and under the pressing processes of globalization, bioethics has rapidly evolved into a strong discipline with a wider conceptual and methodological framework.

This type of bioethics remains in any case subject to the needs of more developed countries, which are confronted with scientific advances and technological innovations. From here stem a bioethical framework that is often irrelevant for the majority of the world population living in less developed regions, with limited or no access to health care and benefits from the progress in science and technology.

Supranational bioethics has therefore emerged as a new type of discourse, specifically devoted to the impact of globalization on citizens across the world. The traditional approaches on advanced technologies, scientific researches, and sophisticated health care are no longer sufficient. Bioethics needs to be extended, taking into account the practical effects of globalization and focusing on the forgotten, invisible, ignored billions of people, who are powerless and voiceless Gostin and Dhai 2012: 33–37.

Together with the human dignity and human rights discourse, supranational forum have in particular been trying to offer an innovative platform for redefining crucial notions of bioethics and, in this perspective, stimulating national legislators to harmonize their laws with those of the others States. This also implies the need to find a balance between 'unity and 'diversity', that is the peaceful coexistence between several socio-cultural and legal traditions in an increasingly globalized viewpoint. The major contribution of UNESCO over the two past decades is that it has contributed to this change, promoting a broader view of bioethics, which is more and more involved in human dignity and human rights concerns. The 2005 Declaration could be considered as a result of this effort. The challenge is now to put its principles into practice.

We should in fact be aware that, while it is of crucial importance, until now the 2005 document is only the first stage towards a common and better understanding of current bioethical issues. We clearly need to make a step forward.

References

Alicino, Francesco 2010, Constitutionalism as a Peaceful "Site" of Religious Struggles, *Global Jurist*, 8: 1–32.

Andorno, Roberto 2009, Human Dignity and Human Rights as a Common Ground for a Global Bioethics, *Journal of Medicine and Philosophy*, 34: 223–240.

Boussard, Hélène 2009, Article 22: Role of States, in *The UNESCO Universal Declaration on Bioethics and Human Rights: Background, Principles and Application* (pp. 293–303) ten Have, Henk, Jean, Michèle S., eds., Paris: UNESCO.

Council of Europe 1997, *Convention for the Protection of Human Rights and Dignity of the Human Being with Regard to the Application of Biology and Medicine: Convention on Human Rights and Biomedicine*, European Treaty Series – No. 164.

Council of the European Union 2016, *Position of the Council at First Reading with a View to the Adoption of a Regulation of the European Parliament and of the Council on the Protection of Natural Persons with Regard to the Processing of Personal Data and on the Free Movement of Such Data, and Repealing. Directive 95/46/Ec* (General Data Protection Regulation), St 5419 2016 Init – 2012/011 (Olp).

European Court of Human Rights (Fourth Section) 2006, *Case of Evans v. The United Kingdom*. Application No. 6339/05.

European Data Protection Supervisor 2015, *Opinion 4/2015 Towards a New Digital Ethics Data, Dignity and Technology*, <https://edps.europa.eu/sites/edp/.../15-09-11_data_ethics_en.pdf> last accessed 02-08-17.

Farmer, Paul 2003, *Pathologies of Power. Health, Human Rights, and the New War on the Poor*, Berkeley: University of California Press.

Floridi, Luciano 2016, *On Human Dignity as a Foundation for the Right to Privacy*, Philosophy & Technology, <www.consorzio-cini.it/index.php/en/component/.../579> last accessed 02-08-2017.

Gert, Bernard 2014, A Global Ethical Framework for Bioethics, in *Global Bioethics and Human Rights Contemporary Issues* (pp. 12–27) Teays, Wanda, Gordon, John-Stewart, Renteln, Alison Dundeln eds., Lanham, Boulder, New York, Toronto, Plymouth, UK: Rowman & Littlefield.

Gostin, Lawrence O., Dhai, Ames 2012, Global Health Justice: A Perspective from the Global South on a Framework Convention on Global Health, *South African Journal of Bioethics & Law*, 5: 33–37.

HUGO 1996, *Statement on the Principled Conduct of Genetics Research*, <https://www1.umn.edu/humanrts/instree/geneticsresearch.html> last accessed 26-05-2016.

IBC 2008, *Report of the International Bioethics Committee of UNESCO on Consent*, Paris: UNESCO.

IBC 2010, *Report of the International Bioethics Committee of UNESCO on Social Responsibility and Health*, Paris: UNESCO.

IBC 2013, *The Principle of Respect for Human Vulnerability and Personal Integrity: Report of the International Bioethics Committee of UNESCO (IBC)*, Paris: UNESCO.

IBC 2013b, *Report on Traditional Medicine Systems and Their Ethical Implications*, Paris: UNESCO, <http://unesdoc.unesco.org/images/0021/002174/217457e.pdf> last accessed 28-05-2016.

Kirby, Michael 2005, *UNESCO and Universal Principles on Bioethics: What's Next? IBC, Twelfth Session of the International Bioethics Committee (IBC)*, Tokyo, Japan, 15–17 December 2005. Abstracts or Texts of the Presentations of the Speakers. Paris: UNESCO <www.hcourt.gov.au/assets/publications/speeches/former-justices/kirbyj/kirbyj_18dec05.pdf> last accessed 28-05-2016.

Kirby, Michael 2008–2009, Human Rights and Bioethics: The Universal Declaration of Human Rights and UNESCO Universal Declaration of Bioethics and Human Rights, *Journal of Contemporary Health Law and Policy*, 25: 309–331.

Levitt, Mairi, Zwart, Hub 2009, Bioethics: An Export Product? Reflections on Hands-on Involvement in Exploring the "External" Validity of International Bioethical Declarations, *Bioethical Inquiry*, 6: 367–377.

Macpherson, Cheryl Cox 2007, Global Bioethics: Did the Universal Declaration on Bioethics and Human Rights Miss the Boat?, *Journal of Medical Ethics*, 33: 588–590.

Magnus, Richard 2016, The Universality of the UNESCO Universal Declaration on Bioethics and Human Rights, in *Global Bioethics: The Impact of the UNESCO International Bioethics Committee* (pp. 29–42) Bagheri, Alireza et al., eds. Basel: Springer AG.

Office of the High Commissioner for Human Rights 2003, *Commission on Human Rights Resolution 2003/69 Human Rights and Bioethics*. 62nd Meeting 25 April 2003, Geneva, Chap. XVII., E/CN.4/2003/L.11/Add.7.

Pavone, Ilja Richard 2009, *La Convenzione europea sulla biomedicina*, Giuffré: Milano.

Pavone, Ilja Richard 2012, Medical Research in Developing Countries and Human Rights, in *Human Medical Research: Ethical, Legal and Cultural Aspects* (pp. 65–87), Schildmann, Jan, Sandow, Verena, Rauprich, Oliver, Vollmann Jochen, eds., Basel: Springer AG.

Plomer, Aurora 2005, *The Law and Ethics of Medical Research. International Bioethics and Human Rights*, London: Cavendish Publishing Limited.

Schmidt, Harald 2007, Whose Dignity? Resolving Ambiguities in the Scope of "Human Dignity" in the Universal Declaration on Bioethics and Human Rights, *Journal of Medical Ethics*, 33: 578–584.

Schuklenk, Udo 2010, Defending the Indefensible: The UNESCO Declaration on Bioethics and Human Rights: A Reply to Levitt and Zwart, *Bioethical Inquiry*, 7: 83–87.

Solinís, Germán (ed.) 2015, *Global Bioethics: What for? Twentieth Anniversary of UNESCO's Bioethics Programme*, Paris: UNESCO.

Somit, Albert, Peterson, A. Steven 2012, Biopolicy: A Critical Linkage, in *Biopolicy: The Life Sciences and Public Policy* (pp. 3–12) Somit, Albert, Peterson, A. Steven, eds., Bingley: Emerald Group Publishing Limited.

Stanton-Jean, Michèle 2016, The UNESCO Universal Declarations: Paperwork or Added Value to the International Conversation on Bioethics? The Example of the Universal Declaration on Bioethics and Human Rights, in *Global Bioethics: The Impact of the UNESCO International Bioethics Committee* (5: pp. 13–22) Bagheri Alireza et al. ed., Basel: Springer AG.

Stout, Jeffrey 2004, *Democracy and Tradition*, Princeton: Princeton University Press.

UNESCO 2004, *Resolution 15 Adopted by the General Conference at its 32nd Session*, Paris, 29 September to 17 October 2003, Paris: UNESCO.

UNESCO 2009, *The UNESCO Universal Declaration on Bioethics and Human Rights: Background, Principles and Applications*, Paris: UNESCO.

UNESCO 2010, *National Bioethics Committees in Action*, Paris: UNESCO.

UNESCO 2011, *Casebook on Human Dignity and Human Rights, Bioethics Core Curriculum Casebook Series*, Paris: UNESCO.

Wolinsky, Howard 2006, Bioethics for the World, *EMBO Reports*, 7: 354–358.

World Medical Association (WMA) 1964, *Declaration of Helsinki – Ethical Principles for Medical Research Involving Human Subjects, as Amended by the 64th WMA General Assembly*, Fortaleza, Brazil, October 2013, <www.wma.net/policies-post/wma-declaration-of-helsinki-ethical-principles-for-medical-research-involving-human-subjects/> last accessed 02-08-2017.

VIII The role of soft law in bioethics

Ilja Richard Pavone

Introduction

UNESCO is the only UN Specialized Agency with a specific mandate on bioethics and life sciences, and it has accordingly established international standards on the protection of human rights in the domain of science and medicine (Köch and Fischer 1997: 455). Mention should be made of the three universal declarations adopted by the General Conference of UNESCO, which constitute the pillar of 'international biolaw': the Universal Declaration on the Human Genome and Human Rights (UDHGHR, 1997), the International Declaration on Human Genetic Data (IDHGD, 2003), the Universal Declaration on Bioethics and Human Rights (UDBHR, 2005). The UDBHR was the culmination of an original design aimed at including bioethics within international human rights law and to give rise to the 'universal law of bioethics' (Kirby 2009: 326). It had the goal of promoting common and shared principles and shaping, using the words of Baker, a "negotiated moral order", which represents a compromise between moral areas open to a negotiation and areas closed to any negotiation (Baker 1998: 233).

In its standard-setting activity in this field, UNESCO opted in favour of soft law instruments, or non-binding agreements rather than hard law instruments. The question arises as to why UNESCO opted for this choice. The answer isn't quite as easy, since the achievement of a consensus on a binding treaty on these so divisive matters would have been a 'mission impossible' (Anglois 2013: 65). Just consider the difficulties met at the regional level to approve the text of the Council of European Convention on Human Rights and Biomedicine (Biomedicine Convention or Oviedo Convention), which was eventually adopted without the participation of important European States, like Germany and United Kingdom, that (for opposite reasons) did not sign the treaty (Birnbacher 2001; Pavone 2009: 44).[1] Therefore, representative of governments realized during the negotiation process of the UDBHR that too rigidly defined duties would have only led to inefficiency, by deterring States from signing or ratifying a potential convention or treaty, or approving a declaration.

The goal of this chapter is to argue that non-binding instruments, far from being purely rhetorical as some may believe, are – under a realist view of international law (Kranser 2002) – the most suitable legal tool in order to come to terms with the concepts and

---

1 Germany criticized the possibility envisaged by the Convention to carry out non-therapeutic research on incapable persons (considering therefore the Convention as "too indulgent" towards scientific research). United Kingdom, instead, blamed the prohibition to create embryos in vitro for research purposes established pursuant to Article 18, Para. 2, of the Convention (considering therefore the Convention as "too restrictive" towards scientific research) (Birnbacher 2001: 461).

categories related to global bioethics and to fill the legal vacuum on this subject. In addition, I will sustain that those instruments are the first step in the pursuit of a global common framework on bioethics, through the passage from a liberal view to a communitarian ideology, promoting a more salient role for concepts of solidarity, community and public interest. Any attempt to negotiate a universal treaty on these matters, on the model of the core human rights treaties would be, therefore, doomed to failure.

Soft law in the sources of international law

As affirmed by Raustiala, international agreements can be negotiated as binding as well as non-binding agreements (the author differentiates between *contracts* and *pledges*, whereas the first create legal obligations and the latter only moral or political commitments) (Raustiala 2005: 581).

The term 'soft law' – coined by the Anglo-American doctrine – is generally used to indicate a series of acts, located in a 'grey area' of international law, in contrast with the 'white area' of hard law, not homogenous as to their origin and nature that, although void of binding legal effects, can however have a legal relevance in the long term (the only immediate effect is in the field of good faith). The authors of the 'classical school' of international law, however, deny any legal effect to soft law instruments (Gross 1965: 48).

Indeed, soft law is deemed to establish and delineate objectives to be achieved in the long term rather than in the present, programs rather than prescriptions, guidelines rather than strict obligations.

Soft law rules are characterized by elasticity, flexibility and vagueness of the contents and of the scopes: in fact, the UDBHR aims to "provide a universal framework of principles and procedures to guide States in the formulation of their legislation, policies or other instruments in the field of bioethics" (Article 2 Paragraph 1).

This terminology is meant to indicate that the instrument or provision in question is not in itself 'law', but its importance within the general framework of international legal development is such that particular attention requires to be paid to it (Shaw 2014: 83–84).

Although a univocal definition does not exist, soft law can however be described as "normative provisions contained in non-binding texts" (Shelton 2003: 292). This phenomenon has been so qualified by scholars, because of the easy and flexible method of law creation;[2] in fact soft law offers a simpler package to accept by States than hard law (for a critical overview, see Klabbers 1995: 167).

Soft law is not included within the 'traditional' sources of international law listed in Article 38 of the Statute of the International Court of Justice (ICJ) that are treaties and customary law.[3] According to the Vienna Convention on the Law of the Treaties (1969), "treaty means an international agreement concluded between States in written form and governed by international law, whether embodied in a single instrument or in two or more related instruments and whatever its particular designation" (Article 2.1, a). Customary international law "results from a general and consistent practice of states followed by them

---

2 Report of the Secretary General on international legal instruments and mechanism, UN Doc. E/CN.17/1996/17/add 1, page 12.
3 Other sources listed in Article 38 of the ICJ Statute are 'General Principles of Law', 'Judicial Decisions' and 'Writings'.

from a sense of legal obligation".[4] In fact, according to a 'classical view', sources are "the sum of the substantive rules, principles or other materials from which a particular legal norms is nourished" (Friedmann 1963: 279).

Soft law encompasses a broad range of acts, *inter alia*, resolutions of intergovernmental organizations – IGOs (in particular resolutions of the UN General Assembly), such as the Universal Declaration of Human Rights (1948) and the Declaration on the Principles of Friendly Relations Among Nations (1970),[5] declarations adopted as the outcome of international conferences convened by the United Nations, such as the Stockholm Declaration on Human Environment (1972), the Rio Declaration on Environment and Development (1992), the Agenda 21 (1992), the Millennium Development Goals (2000), the Johannesburg Declaration on Sustainable Development (2002) and the Sustainable Development Goals (2015); declarations adopted as the outcome of high-level political meetings, such as the Conference on Security and Cooperation in Europe (CSCE) Helsinki Final Act, that led to the development of the human rights doctrine in Eastern Europe; resolutions dealing with topics that traditionally fall outside States' domestic jurisdiction, such as disarmament, outer space, the deep seabed, marine protection. In the field of fisheries management one can register, for instance, several codes of conduct, guidelines and plans of actions of IGOs and UN Specialized Agencies. Examples of important acts on this topic are the UN General Assembly Fisheries Resolutions, the FAO Code of Conduct for Responsible Fisheries, the FAO International Code of Conduct on Pesticide Management, the FAO International Plan of Action for Conservation and Management of Sharks (Ipsoa-Sharks) and the Memorandum of Understanding on the Conservation of Migratory Sharks adopted in 2010 under the auspices of the Convention on Conservation of Migratory Species of Wild Animals (CMS).

Environmental law is an example of a branch of international law where, as observed by Heriksen, soft law instruments are "prevalent" (Heriksen 2017: 37), starting with the Stockholm Conference, whose core principles have been then 'codified' by domestic legislations and conventions, treaties and protocols. Some of them, through the spontaneous observance by States, have over time acquired the status of customary rules (i.e. the precautionary principle, the no-harm principle, the polluter-pays principle, the duty of cooperation, notification and consultation) (Birnie, Boyle and Redgwell, 2009: 12).

Another domain characterized by a proliferation of soft law instruments is that of emerging technologies, which comprehends nanotechnology and cyber security (Marchant and Allenby 2017: 108). In this area, for instance, the International Organization for Standardization (ISO) set international standards for nanotechnology risk management (2009), and the Organization of American States (OSA) adopted the Declaration Strengthening Cybersecurity in the Americas (2012).

In this context, mention should also be made of standards developed by the World Organization for Animal Health (former *Office International des Epizooties* – OIE) in the emerging field of animal welfare law, such as the Terrestrial Animal Health Code and the Aquatic Animal Health Code.

---

4 American Law Institute, *Restatement of the Law, Third, the Foreign Relations Law of the United States,* American Law Institute Publishers, St. Paul, MN 1987. §102(2).

5 GA Declarations, in particular, can be defined, according to the UN practice, as a solemn instrument resorted to only in very rare cases relating to matters of major and lasting importance where maximum compliance is expected (Report of the Commission on Human Rights, United Nations document E/3616/Rev. l, paragraph 105, eighteenth session, Economic and Social Council, 19 March–14 April 1962, United Nations, New York).

When trying to establish the specific legal relevance of soft law and the distinction between mere social rules and law, one aspect emerges. Soft law is supposed to have legal effects in the long term only if and when States comply with a determined set of rules enshrined in a declaration: while treaties pose immediately duties of implementation upon States Parties after their entry into force, soft law norms are only "*potentially* binding" (Andorno 2013b: 69). Soft law is indeed conceived "as the *beginning of a gradual process*" which requires further steps to make such acts compulsory for States; otherwise, it would not be labeled as 'law' but merely as ethics or moral principles (or at least, I would add as a social norm) (Andorno 2007).[6]

In the process of 'hardening', soft law firstly permits and facilitates the assessment of customary rules or general principles of international law, performing a function of interpretation of international law. Second, it can be a first step in the creation of non-written rules, since it permits and facilitates the appraisal of customary norm. Third, it is sometimes the beginning of a treaty-making process, paving the way to the adoption of a binding treaty, as in the case of the Universal Declaration on Human Rights, whose content had been then translated in the Covenant on Civil and Political Rights (1966) and in the Covenant on Economic, Social and Cultural Rights (1966).

In the long term, the principles enshrined in the UDBHR did not however create hard law by leading to the adoption of a convention. However, scholars assumed that for instance the Nuremberg Code, a soft law instrument, gained the status of "international legal document" (Annas 1992: 121). Other scholars argue that the principles enshrined in the Nuremberg Code, the Declaration of Helsinki and CIOMS Guidelines have become customary international law binding on all States except persistent objectors (Fidler 2001: 326). The same happened with the Declaration; it did not create new norms, but rather codified and reiterated existing principles in the domain of bioethics.

The role of declarations in UNESCO standard-setting activity

A declaration is defined as "a solemn instrument resorted to only in very rare cases relating to matters of major and lasting importance where maximum compliance is expected".[7] As previously examined, declarations do not have any binding legal effect, although, in the IGOs' practice, some of them can constitute the material source of legal rights and duties. This is the case, for instance, of an interpretative declaration, an instrument annexed to a convention, with the aim of providing an interpretation to its terms (Asamoah 1966).[8] Another case is provided by the principles contained in resolutions of international conferences that

---

6 The questions on the value of soft law had been raised when the UDHR was adopted in 1948. In this regard, the International Court of Justice (ICJ), in its advisory opinion on the Legality of the Threat or Use of Nuclear Weapons, notes that General Assembly resolutions, even if they are not binding, may sometimes have normative value.

7 Report of the Commission on Human Rights, United Nations document E/3616/Rev. 1, paragraph 105, eighteenth session, Economic and Social Council, 19 March–14 April 1962, United Nations, New York.

8 A similar process has been endorsed at the regional level by the Council of Europe, which sustains its treaties with explanatory reports that provide information to clarify the object and purpose of a convention and to better understand the scope of its provisions. See, for instance, the Explanatory Report to the Convention for the Protection of Human Rights and Dignity of the Human Being with Regard to the Application of Biology and Medicine, ETS No. 164, available at https://rm.coe.int/CoERMPublicCommonSearchServices/DisplayDCTMContent?documentId=09000016800ccde5.

can be 'translated' in an international treaty, as it happened and still happens in the sector of international environmental law; or they can be recalled in a treaty, whereas it explicitly establishes that States Parties oblige themselves to respect the provisions of the soft law act recalled. For instance, the Polish–German Treaty of Good Neighbourship and Friendly Cooperation (1991) establishes that the Parties commit themselves to respect the principles established by CSCE on minorities.

As stated by Sloan, declarations adopted by the General Assembly are binding as they emanate from a body that operates as an "agent of the international community" (Sloan 1948: 19).

It is worth underlying, in this regard, that under a legal point of view, there is not a real difference between a recommendation and a declaration in the practice of the United Nations and its Specialized Agencies; on the contrary, the UN General Assembly often adopts recommendations that do contain declarations (Marchisio 2012: 163).

The same discourse can be translated to UNESCO practice; although there is not an explicit provision in the UNESCO Constitution mentioning declarations,[9] this has not, however, diminished their role in the standard-setting activity of the Organization (Yusuf 2006: 130). In fact, in its practice, the UNESCO General Conference adopted several declarations; this kind of legal instrument became, in fact, quite common in the General Conference activity since 1966, as the first UNESCO declaration was adopted (Declaration on Principles of International Cultural Cooperation). In order to 'codify' its standard-setting activity through declarations, the General Conference enacted a legal framework for the elaboration, examination, adoption and follow-up of declarations, charters and similar standard-setting instruments (Resolution 33 C).[10] According to the procedural rules, the adoption of declarations "do not rest on any specific constitutional or regulatory basis but draw on practice within the Organization, in contrast to conventions and recommendations". Declarations are generally adopted by consensus, while conventions and recommendations require respectively a two-thirds majority and a simple majority within the General Conference. We can affirm that the UNESCO General Conference expresses its legislative powers (*de lege lata*) through the adoption of conventions as well as of declarations.

The difficult struggle for consensus in the field of bioethics

The UDBHR was the peak of the standard-setting activity of UNESCO in the field of bioethics, determining the establishment of a corpus of international rules, labeled as

---

9 UNESCO, since its establishment in 1946, adopted several legal instruments in the field of education, science, culture, according to Article IV.B.4 of the its Statute ('standard-setting instruments'), which attributes a normative function to the General Conference. Two categories of legal instruments – conventions and recommendation, unlike declarations – are specifically mentioned in Article IV.B.4, and can be considered as the main standard-setting instruments in the promotion of the goals of the Organization. The first convention was the Agreement for Facilitating the International Circulation of Visual and Auditory Materials of an Education, Scientific and Cultural Character (1948); the first recommendation was the Recommendation on International Principles Applicable to Archeological Excavations (1956).

10 The procedural rules for the adoption of a declaration have been established during the 33rd session of the General Conference. (Multi-stage procedure for the elaboration, examination, adoption and follow-up of declarations, charters and similar standard-setting instruments adopted by the General Conference). See Legal framework for the elaboration, examination, adoption and follow-up of declarations, charters and other similar standard-setting instruments (33 C/20), Paragraph 3, available online at http://portal.unesco.org/en/ev.php-URL_ID=28555&URL_DO=DO_TOPIC&URL_SECTION=201.html.

"international biolaw" (Andorno 2013b). It is a branch of international human rights law of recent evolution, which had its normative peak between 1997 and 2005, as part of the "expansive phase of bioethics" (Caporale 2015: 219), and now is facing a 'crisis' in the crafting of new norms, at least at the global level.[11] The same discourse can be translated to the sector of environment, whereas after a period of intense standard-setting through the adoption of binding treaties in the phase of 'environmental globalism', such as the Framework Convention on Climate Change (1992), the Biological Diversity Convention (1992), the Convention on Desertification (1994), the law-making process registered few legal developments the last decade.

The most relevant novelty in this field is represented by the Paris Agreement on climate change of 2015 (into force since 4 November 2016). This Agreement however does not contemplate any concrete commitment upon States, limiting its scope to general obligations, and it is based upon "mandatory and non-mandatory provisions relating to parties' mitigation contributions" (Bodansky 2016: 142).[12]

The new branch of international law named as biolaw is characterized by a relative exiguity of binding international agreements and at the same time by a proliferation of soft law rules, in what can be considered as an excessive fragmented and pluralistic regulation. On the international stage, the three 'core bioethical texts' (the Declaration of Geneva, the Nuremberg Code, the International Code of Medical Ethics), whose principles are nowadays widely accepted and followed by the scientific community, are soft law agreements. To these acts one may add the WMA (World Medical Association) Declaration of Helsinki (DOH [and its revised versions]), the CIOMS (Council for International Organizations of Medical Sciences) Ethical Guidelines on Biomedical Research, the two WHO resolutions on human cloning,[13] the WHO World Health Regulations, the UN Declaration on Human Cloning, the ECOSOC Resolution 2004/22 on Preventing, Combating and Punishing Traffic in Organs, the above-mentioned UNESCO declarations. At the regional level, the Council of Europe since 1976 adopted several recommendations and resolutions on bioethical matters, ranging from xenotransplantations to biobanks.[14] In the American continent, the Pan American Health Organization enacted in 2012 a Resolution on bioethics and the integration of health in bioethics, in which the Organization quoted, *inter alia*, the UDBHR.[15]

---

11 At the regional level, one can mention, as a legal development, the EU Regulation No. 536/2014 on Clinical Trials on Medicinal Products for Human Use, or the CoE Convention on Action against Trafficking in Human Beings (2015).

12 In the United States, the former president Barack Obama could directly ratify the Agreement (later challenged by Donald J. Trump), bypassing the Senate, since the climate deal was considered not as a 'Treaty', but as a mere 'Executive Agreement'. 'President Obama: The United States Formally Enters the Paris Agreement', <https://obamawhitehouse.archives.gov/blog/2016/09/03/president-obama-united-states-formally-enters-paris-agreement> last accessed 02-08-2017.

13 Resolution WHA 50.37 on Human Cloning for Reproductive Purposes and Resolution WHA 51.10 on Ethical, Scientific and Social Implications of Cloning in Human Health.

14 See, for instance, Recommendation Rec(2003)10 of the Committee of Ministers to Member States on xenotransplantation Recommendation Rec(2006)4 of the Committee of Ministers to Member States on research on biological materials of human origin.

15 28.a Conferencia Sanitaria Panamericana, 17–21 September 2012, Resolución CSP28.R18, Bioética: Hacia la integración de la ética en el ámbito de la salud.

Also at the domestic level, the regulation of some branches of science and medicine is still carried out through soft law instruments, such as ministerial guidelines,[16] codes of conduct,[17] recommendations, deontological codes, without resorting to laws, decrees or other 'hard' legal instruments.[18]

The adoption of treaties relies entirely, in fact, on the political will of States, and their implementation is determined by their suitability to the domestic social and political setting. The case of the United States and climate change is emblematic in this regard. The Obama administration pushed for the adoption of a new climate change deal in substitution of the Kyoto Protocol and was at the forefront in climate change negotiations; instead, President Obama's successor Donald Trump, due to a different view on global warming and his wish to protect first US domestic interests, decided to withdraw the United States from the Agreement.[19]

In the light of the above-mentioned central role of national interests in making commitments at the international level, it is not surprising therefore, that the only treaties on such a sensitive issue as bioethics have been enacted at the regional level by the Council of Europe, with the adoption of the Biomedicine Convention and its four additional protocols.[20] In that case, the CoE's Council of Ministers decided to adopt a framework convention with the aim of 'codifying' the consistent amount of recommendations on bioethical matters adopted since 1976. In this regard, it is worth recalling also the activity of the European Union (EU) in the field of bioethics, through the inclusion of core principles in Article 3 of the Charter on Fundamental Rights (Right to Integrity of the Person), such as the respect for informed consent and the prohibition of practices that imply a financial gain of the human body (surrogate motherhood or organ trafficking), and the banning of human cloning. The Charter was initially void of any legal effect, before being incorporated into the Treaty on the European Union (TEU, Lisbon version, into force since 2009).

More in general terms, the choice to enact just soft law rules in the field of global bioethics is in line with the stagnation in terms of quantitative and qualitative production of "formal international law" in favour of "informal international lawmaking" (Pauwelyn, Wessel and Wouters 2014: 734).

Several scholars criticized the fact that the existing international rules on bioethics lack legal enforceability and are often vaguely worded (Macpherson 2007: 588). In the

---

16 In Italy, many controversial aspects of Law No. 40/2004 (Rules on medically assisted procreation) have been at a later time clarified or modified not by a new law, but by ministerial decrees and guidelines adopted by the Health Ministry (often following judicial judgements).

17 One can quote as example in Italy, the Code of Ethics of the Italian National Institute of Health.

18 Municipal law today operates in the context of international law. Principles expressed even in a declaration of principles can have an impact at the domestic level (i.e. judicial decisions). For instance, the European Court of Human Rights, in the case *Evans v. the United Kingdom* (Application No. 6339/05, Judgement of 10 April 2007) quoted Article 6 of the UDBHR (Para. 52) – amongst the relevant international texts – in support of its thesis. It does not mean that, as such, the principles of the UDBHR bind a judge or a State to their implementation; it simply means that they provide an interpretative tool to the judge that must apply international shared standards.

19 Statement by President Trump on the Paris Climate Accord, 1 June 2017, https://www.whitehouse.gov/the-press-office/2017/06/01/statement-president-trump-paris-climate-accord.

20 Additional Protocol on the Prohibition of Cloning Human Beings (CETS No. 168); Additional Protocol Concerning Transplantation of Organs and Tissues of Human Origin (CETS No. 186); Additional Protocol Concerning Biomedical Research (CETS No. 195); Additional Protocol Concerning Genetic Testing for Health Purposes (CETS No. 203).

particular case of the UDBHR, it was not an easy task to enucleate bioethics within the human rights discourse; indeed, from a European perspective, the link between international human rights law, human rights and bioethics seems obvious, but not everybody fully agrees to this stance in other countries (Ashcroft 2010: 639). For instance, the concept of 'human dignity' is not the same throughout the world and differs according to factors such as culture, nation and religion. In fact, some non-European delegates were concerned about the European footprint of the basic values and principles that have been established in the Declaration. In this regard, Western Europe and North American States on one side, and Asian States on the other side, were at odds over whether the Declaration should be based on a human rights framework. The Declaration eventually reconciled these different views, enucleating "basic" of "fundamental" principles universally accepted, and shaping a negotiated moral order (Baker 1998: 233) on some topics which formed the negotiation platform amongst the different stakeholders involved in the drafting process.

The price to be paid for the global endorsement of the few principles enshrined in the UDBHR consisted in the impossibility to craft more detailed provisions in a binding treaty. However, the UNESCO Declaration on Bioethics must not be confused with other guidelines and declarations devised by non-governmental professional organizations, such as WMA or CIOMS, that can be defined as purely political arrangements or codes of professional ethics. In fact, these set of codes of conduct and guidelines adopted by NGOs, like the Declaration of Helsinki (DOH), cannot be classified as soft law. The legal value of the DOH – for instance – lays primarily in the influence it had over decades on the medical class in the development of international and national codes of conduct and domestic rules that incorporated its principles (Plomer 2005: 5). The DOH is not however a legal rule *stricto sensu*, despite its strong impact at the domestic level and its influence on the shaping of municipal law on biomedical research. Indeed, as argued by Thürer, only subjects of international law (states and IGOs) can adopt legal instruments classified as soft law (Thürer 2012: 271). UNESCO declarations have been, in fact, devised by an intergovernmental organization, which represent the will of the governments and have therefore a higher hierarchical value than NGOs' codes of conduct and guidelines. In this regard, it is important to underline that the same term 'law' encompasses legal effects; otherwise, one would use another term instead of 'soft law', i.e. code of ethics, moral code, code of conduct, guideline. Andorno clearly points out that UNESCO declarations, given their intergovernmental nature, cannot be classified as "purely ethical or rhetorical recommendations deprived of any legal effect" (Andorno 2009: 225).

The choice of soft law in negotiating global bioethics as the only option

As previously observed, the standard-setting activity of UNESCO in the field of bioethics relied upon the legal instrument of the 'declaration' of an international organization.

The UDBHR represents a global consensus on some bioethical topics, such as the recognition of the moral supremacy of human beings over the advancements of science on the basis of the concept of human dignity and of the principles of informed consent, individual autonomy and self-determination as human rights, the reaffirmation of the protection of future generations, as well as of the environment, the biosphere and biodiversity (Magnus 2016: 29).

The adoption of the UDBHR posed and still poses two main problems strictly interrelated, concerning respectively its content and its legal value. As to the first point, many important bioethical issues have not been included in the Declaration, because they are too divisive and the moral values at stake and the religious beliefs were too different. The final version of the Declaration did not eventually include non-negotiable issues as broad-ranging as the beginning of life (access to artificially procreative techniques, procreative rights, the legal status of the human embryo, the legitimacy of scientific research on human embryos, stem cell research, gene therapy) and the end of life (active or passive euthanasia, assisted suicide), included the right to access to palliative care or the question of the legal value of the advanced directives. These shortcomings have been the object of criticism (Benatar 2005: 220–224; Macpherson 2007: 588–590), highlighting the failure of the UDBHR to introduce new bioethical principles (that should have then been translated into duties). In particular, other scholars underlined the fact that the Declaration simply restated widely accepted principles, such as informed consent and individual autonomy, which are already contained in several municipal laws or domestic guidelines or codes of conduct, just listing desirable objectives, without establishing concrete measures to implement them (Williams 2005: 210). Technology transfer and benefit sharing are amongst the few innovative principles established by the Declaration, while no mention is made to big data and health (Faunce and Nasu 2009). However, Andorno argued in favour of the Declaration, stating that "the greatest merit of this instrument is to gather those principles and to integrate them into a human rights framework. In sum, the purpose of the declaration is not to invent new bioethical principles or to provide the definitive solution to the growing list of bioethical dilemmas" (Andorno 2013b: 84).

Despite these shortcomings, its adoption represented nonetheless a first but significant step in the pursuit of common shared values in the field of bioethics and in the creation of customary law. In this regard, the simple fact that all the States of the international community (the Declaration was adopted by consensus) agreed on common bioethical principles is in itself an accomplishment in the promotion of human rights in the domain of medicine. On this subject, ten Have affirmed that: "The unanimous adoption by the member States is not merely symbolic but gives the declaration moral authority and creates a moral commitment" (ten Have 2011: 21).

As to the issue of the legal value of this global instrument on bioethics, in the light of the difficulty of reaching an agreement on some ethical issues because of being too divisive, the adoption of a strictly treaty-based instrument such as a framework convention – as previously observed – would not have been a feasible option (although the initial proposal by the French delegation provided for the elaboration of a convention on bioethics) (Idrissi 2009: 317). It is not surprising, therefore, that the International Bioethics Committee (IBC), in its preparatory report of 2003, purposely recommended the choice of a non-binding agreement, dropping its initial idea of drafting a convention. In particular, it affirmed that "given that the aim of such an instrument will by its nature be broad and will receive the broadest acceptance possible by public authorities, the scientific community and the general public, the Working Group considers it preferable, in the initial stage, to settle on a declaration".[21]

In its 2003 report the Working Group went further, and in support of its thesis, recalled at Para. 43 cases of UN treaties in the field of human rights that have been preceded by a declaration (the two International Covenants of 1966, the Convention on the Elimination

---

21 'Report of the IBC on the Possibility of Elaborating a Universal Instrument on Bioethics', SHS/EST/02/CIB-9/5 (Rev. 3), (Rapporteurs: Leonardo de Castro and Giovanni Berlinguer), 13 June 2003, paragraph 42; Available at: http://unesdoc.unesco.org/images/0013/001302/130223e.pdf.

of All Forms of Racial Discrimination, the Convention on the Elimination of All Forms of Discrimination Against Women and the Convention on the Rights of the Child).[22] In fact, the UN negotiation practice in the field of human rights has sometimes foreseen, as a first step, the adoption of a declaration which contains guidelines and an invitation to States to follow them, which generally anticipates, as a second step, the adoption of a treaty or convention on the same topic.

Also in the cases of the UDHGHR and of the IDHGD, the IBC opted for soft law instruments, providing several justifications. For instance, the IBC affirmed that it would have been easier and quicker to find an agreement on non-binding instruments that, therefore, do not request 'formal steps' such ratification, accession or acceptance (Langlois 2013: 65).

The Ad Hoc Committee on Human Cloning confronted itself with the same problems in negotiating a binding agreement with the aim of regulating this technique.[23] In this case, a Working Group of the Sixth Legal Committee of the GA started to work on an international convention banning human cloning, but it was eventually forced to rely on a declaration (Pavone 2008: 3). The reason for this failure was due to the irreconcilable differences on the interpretation of the term 'human cloning'. In this regard, a group of States led by Costa Rica and the United States wanted to extend the ban of reproductive human cloning to all kinds of cloning, also including stem cell research ('therapeutic cloning'); while other States, such as Belgium, Canada, United Kingdom, intended to limit the ban to reproductive human cloning. As a compromise, the United Nations Declaration on Human Cloning was eventually adopted on 8 June 2005 (by majority and not by consensus) (Arsanjani 2006: 164).[24] This declaration was however weak since its approval, being that many States (i.e. United Kingdom) voted against its adoption. Its legal value is therefore weakened if compared to the UDBHR, whose framework principles are based on the acceptance of the whole global community. Indeed, a key element that provides some legal value to a declaration is given by the absence of objections (which implies adoption by unanimity or consensus).

The relation between soft law and hard law in the field of bioethics

According to the traditional theory of international law, what differentiates hard law and soft law is the enforceability of treaties and conventions with respect to declaration of principles (Jennings, Watts 1992: 32). In fact, in line with the principle *pacta sunt servanda*, States must comply with the obligations contained in a treaty or convention and must modify their domestic legislations accordingly in order to implement a treaty or convention. In case of violation of their duties of implementation, they commit an internationally wrongful act, entailing an international responsibility. It has legal consequences, as another State Party might, for instance, take countermeasures against the responsible State or bring the case to the attention of the International Court of Justice or of an arbitrate. Treaties

---

22 Another example is represented by the UN Convention on the Rights of Persons of Disabilities (2006), whose forerunner is represented by the UN Standard Rules on the Equalization of Opportunities for Persons with Disabilities (1994).

23 The Committee was established by the GA pursuant to Resolution 56/93 of 28 January 2002 on "International convention against the reproductive cloning of human beings".

24 The Declaration on Human Cloning invites UN Member States "to adopt all measures necessary to prohibit all forms of human cloning inasmuch as they are incompatible with human dignity and the protection of human life" (Annex, lect. *b*).

or conventions generally contain provisions concerning the resolution of disputes that may arise on the interpretation of the treaty or in case of non-compliance. Conversely, the text of the UDBHR (as well as that of other declarations), does not establish any legal consequence in case of breach of its provisions and/or of non-compliance; the UDBHR merely predicts that States should (not 'shall' or 'must') "take all appropriate measures, whether of a legislative, administrative or other character, to give effect to the principles set out in this Declaration" (Article 22, 'Role of States'). Therefore, the drafters of the Declaration invited States "to *promote* rather than to *implement* the content of the UDBHR" (Boussard 2007: 293), since only hard law instruments establish the duty of translating their content in domestic law. As a practical consequence, therefore, no State may invoke the violation of the provisions of the UDBHR by another State before an international court/arbitration, or may adopt countermeasures as a reaction to a wrongful act. In fact, as observed by Dupuy, one major concern in the field of biolaw regards how "to trigger state responsibility for violation of principles of bioethics" (Dupuy 2007: 33).

Therefore, compliance with the standards established by UNESCO declarations – devised in a soft language that is susceptible to subjective evaluation – presents serious problems. In addition, the drafters of the UDBHR failed to introduce a reporting mechanism like that established in the UDHGHR (Boussard 2007: 125). However, one must underline in this regard that, under the UNESCO Constitution, Member States already have a duty of submitting periodical reports to the Organization on the state of implementation of UNESCO instruments,[25] along with an associated follow-up procedure.

The Biomedicine Convention, by contrast, established the duty of the Parties to make available a judicial procedure to prevent or put a stop to an infringement of its norms (Article 23). It therefore covers not only infringements which have already begun and are ongoing, but also the threat of an infringement. Also Article 25 stated that "appropriate sanctions may be applied in event of infringement of a Convention's right". The aim of the sanctions provided for in Article 25 is to guarantee compliance with the provisions of the Convention, although in their practices, States Parties never invoked the mechanism set up by Article 25.

However, despite the formal difference between *duties* or *pledges* and the level of enforceability of hard law if compared to soft law, some scholars have so far argued that the sources of law as expressed in the ICJ Statute are no longer in line with the evolution of the international community (Cárdenas Castañeda 2013: 355). It seems, indeed that, in the light of the transformation of international law, the development and implementation of international law are no longer exclusively reserved to the traditional sources of international law.

It is also interesting to point out that, according to this line of thought, the difference in terms of efficacy between a framework convention and a declaration is not as impressive as it may seem. As stated by Kratochwil "it is highly significant that the hardness or softness of [rules] can no longer be derived simply from the formality or genesis of the instrument" (Kratochwil 1989: 200). Therefore, regardless of formal ratification of a treaty, international agreements and subsequent obligations may also arise, as observed by van Hoof, from a "more formless expression of consent or acceptance" (van Hoof 1983: 181). Is the

---

25 Article VIII ('Reports of Member States') of the UNESCO Constitution, states: "Each Member State shall submit to the Organization, at such times and in such manner as shall be determined by the General Conference, reports on the laws, regulations and statistics relating to its educational, scientific and cultural institutions and activities, and on the action taken upon the recommendations and conventions referred to in Article IV, paragraph 4".

will of the States and the level of detail of an international act that determines the legal nature of an agreement. In some cases, we can find less detailed norms in the text of a treaty and more detailed norms in the text of a declaration. As affirmed by Dupuy "the hard or soft nature of the obligation defined in a treaty provision should not necessarily be identified on the sole basis of the formally binding character of the legal instrument in which the concerned norm is integrated and articulated" (Dupuy 1991: 430).

Take for instance the case of 'periodic reports' – that States Parties must submit to a relevant treaty body, explaining on how the provisions of a treaty are being implemented – established as a compliance mechanism both by declarations and treaties (mostly in the field of human rights).[26] In their practice, invariably States are often delayed and overdue in submitting their periodic reports, and in many cases they draw a highly positive picture of their achievements, and deny their failures to comply, without recognizing substantial problems. This happens in particular in the UN system with the human rights treaty regime, demonstrating that despite being in presence of a treaty, the degree of compliance is low (Posner 2014: 72).

Therefore, the real difference between a treaty and a declaration does not rely on its form, but on several factors, such as the language adopted in the act (the more detailed it is and less space it leaves to extensive interpretations), on how much the topic matters for the States and on the consequent role of reputation on that issue, as well as on the compliance measures it established.

Therefore, what is crucial as to the obligatoriness of an international agreement is not its form, but the intention of the Parties as inferred from all the relevant circumstances as to whether they intended to create binding legal relationships between themselves (Barelli 2009: 958). This intention can be 'translated' in a treaty that must be precisely worded and that must specify the exact obligations undertaken or the rights granted. In fact, one can argue that a State decides to comply with the provisions of a treaty or of a declaration not according to its legal value, or because it fears to commit an internationally wrongful act, but in the light of its personal interests to maintain its reputation. I fully agree accordingly with the position of Goldsmith and Posner in *The Limits of International Law*, whereas they claim that international law is too weak to improve the world in any significant way and that it primarily reflects the interests of the most powerful states (Goldsmith and Posner 2007).

In this regard, it is worth recalling the thoughts of some scholars that deem that 'reputation' is the key element that determines compliance with international law by the States, and the formal legal nature of an international rule, whether is a treaty or a declaration, does not have any effect in terms of major or minor compliance by States (although the US withdrawal from the Paris Agreement is an exception to this position). Reputation, as well economic and political interests, are the key elements that determine the behavior of a State in the international arena (Guzman 2006: 379).

Indeed, in State practice, the number of cases that comported the recourse to the instrument of the sanction to enforce a Party to comply with the provisions of a treaty or that have been brought under dispute settlement procedure is very limited. In fact, in general terms, States refrain from imposing sanctions, given the deterioration of bilateral relations that they do imply (Guzman 2002: 1868). In the case of environmental law, the inefficacy

---

26 Examples of committees established by the treaty bodies with the aim of monitoring the core international human rights treaties are the Human Rights Committee, the Committee on the Elimination of Racial Discrimination, the Committee on the Elimination of Discrimination Against Women, the Committee on the Rights of Persons With Disabilities. See www.ohchr.org/EN/HRBodies/Pages/TreatyBodies.aspx, last accessed 02-08-2017.

of traditional sanction mechanisms, determined, the development of a new system, labeled as Non-Compliance Procedures (NCPs) (Fitzmaurice and Redgwell 2000).

More coercive enforcement mechanisms are instead set up by disarmament treaties, that cover topics that are vital to the security of the homeland. Take for instance the Iran Nuclear Deal of 2015,[27] that established a strict compliance procedure. In less vital topics for State security – as in the cases of human rights as well of bioethics – States prefer to envisage less stringent monitoring mechanisms.

It is generally understood that the legal force of soft law corresponds to the *opinio juris* on its content, expressed by the adoption of the text by consensus. As stated by Sohn, "unanimously declarations are a new method of creating customary international law" (Sohn 1978: 22). In this view, some declarations of principles adopted by IGOs, such as the UDBHR, can be compared in terms of compulsoriness to framework conventions. As is well known, framework conventions, unlike lawmaking treaties, are generally vaguely worded and establish weak structures aimed at monitoring the degree of implementation of the treaty (Boczek 2005: 33).

The tendency in lawmaking to produce agreements with an extremely vague content, more declaratory than perceptive, is emblematic of the current phase of stagnation in the international law-making process. In this regard, it is interesting to quote the division made by Raustiala between strong and weak structures that differentiate the degree of enforceability of a treaty (Raustiala 2005: 581). Take for instance the United Nations Framework Convention on Climate Change (1992), which, although formally a treaty, established very general principles that given their non-self-executing nature revealed a scarce practical impact at the domestic level (the same argument can be made with regard to the Paris Agreement on Climate Change of 2015).

More specifically, in the field of bioethics, the failure of the Biomedicine Convention is a mirror image of the impossibility of shaping a global and shared bioethics. Although a regional instrument, the Convention had the aspiration to become a global treaty, being open for signature also to non–Member States of the Council of Europe (like Australia, Canada, Holy See, Japan, Mexico, United States of America), that never signed it. If to this element we add the non-participation of CoE Member States like Austria, Belgium, Germany, Italy and the United Kingdom, the Convention failed in its main purpose of being the basis for the formation of subsequent widely accepted non-written rules in the field of bioethics. In few words, the Oviedo Convention did not follow the path of other international treaties that have codified several branches of international law (i.e. environmental law, law of the sea, space law).

I agree in this regard with the view of Dows, Rocke and Barsoon according to whom the 'legal nature' of an international agreement is determined by the 'depth', which is "the extent to which [an agreement] requires states to depart from what they would have done in its absence" (Dows, Rocke and Barsoon 1996: 383).

The role of the UNESCO standard-setting activity in the formation of non-written norms in the field of bioethics

As stated by Shelton, "non-binding norms can have complex and potentially large impact in the development of international law" (Shelton 2014: 160). Customary law requires

---

27 Joint Comprehensive Plan of Action, Vienna, 14 July 2015, http://eeas.europa.eu/archives/docs/statements-eeas/docs/iran_agreement/iran_joint-comprehensive-plan-of-action_en.pdf.

two elements: *opinio juris* (the psychological element) and *diuturnitas* (State practice). It is generally recognized that whereas the vast majority of States consistently vote for resolutions and declarations on a specific topic, that amounts to *opinio juris* and a binding rule may very well emerge provided that the requisite of *diuturnitas* can be proved (this is not, however, the case of the UN Declaration on Human Cloning, that was adopted by majority). The UDBHR undoubtedly contributed to the emergence of non-written rules in the field of bioethics, evidencing the emergent custom and assisting in establishing the contest of some rules, although its principles had not been uniformly implemented. For instance, in South Africa, the Declaration had little legal impact; in fact, a study carried out by Rheeder, although recognizing the importance of the UDBHR, highlighted a simply 'moral impact' of the UDBHR (Rheeder 2014: 51). In other cases, like that of China (Xiaomei 2009: 5) and Brazil (Cruz, de Lima Torres Oliveira and Cordón Portillo 2010), the Declaration had instead a strong impact and influenced domestic law. However, in many countries the principles set out in the Declaration have been included in professional codes of conduct or guidelines (in Singapore, the Bioethics Advisory Committee included the Declaration amongst its core principles), but not in laws.

The key question is to assess which of them have become part of customary international law. Positive examples are represented by the principle of informed consent; the right to the highest attainable standard of physical and mental health; the right to respect of family and private life; the right to enjoy the benefits of scientific progress and its applications (Smith 2012: 72).

In the case of informed consent, this principle has been affirmed in the UNCCPR (Article 7)[28] and reiterated in all the UNESCO declarations (Article 5, *b* of the UDHGHR; Articles 8 and 9 of the UDHGD; Article 6 of the UDBHR) as well as in the CoE Convention on Biomedicine (Article 5), in its Additional Protocol on Biomedical Research, in the EU Charter on Fundamental Rights (Article 3), in EU directives and regulations,[29] guidelines and codes of conduct of NGOs (i.e. the Declaration of Helsinki), and included in domestic legislations of developing countries.[30] Therefore, the cumulative enunciation of this principle by numerous binding and non-binding instruments helped to express the *opinio juris* of the world community.

It marked the transmutation of the principle of informed consent from a political and moral principle to a legal right and with consequent obligations for States, particularly taken in conjunction with Article 5 of the Oviedo Convention. This principle has been recalled in judgments of international and domestic tribunals.[31]

---

28 Article 7 ("Consent to medical treatment and experimentation") provides that "no one shall be subjected without his free consent to medical or scientific experimentation".

29 See for instance, Directive 2001/20/EC of the European Parliament and of the Council of 4 April 2001 on the approximation of the laws, regulations and administrative provisions of the Member States relating to the implementation of good clinical practice in the conduct of clinical trials on medicinal products for human use, and Regulation (EU) No 536/2014 of the European Parliament and of the Council of 16 April 2014 on clinical trials on medicinal products for human use, and repealing Directive 2001/20/EC (para. 30–33).

30 Informed consent has been, for example, disciplined in the South African National Health Act (2003), in the Tanzanian Guidelines of Ethics for Health Research in Tanzania (2009), and recognized in India by the Supreme Court in the case: *Samira Kohli vs. Prabha Manchanda Dr. & ANR* 1(2008) CPJ 56 (SC).

31 For instance, the Italian Constitutional Court in the Judgment n. 438/2008 established that informed consent is a fundamental right of each individual, which draws its foundation in Article 32

Faunce, in this regard, affirmed that "ethical requirements for informed consent before medical or scientific treatment probably constitute international law as involving general principles of law under article 38 (1) (c) of the *Statute of the International Court of Justice*" (Faunce 2005: 173).

The US Supreme Court went further stating, in particular, that the prohibition against non-consensual human medical experimentation has become a norm of customary international law (*Abdullahi v. Pfizer, Inc.*). The Court has in particular recognized in the requirement of informed consent all the elements of a customary norm, such as *opinio juris* and *diuturnitas* (Annas 2011: 194),[32] quoting, in support of its thesis, several soft law instruments, such as the Nuremberg Code, the Helsinki Declaration and the UDBHR.

This principle has been then included – after the adoption of the UDBHR – in several legislations of non-Western countries, like for instance, China (Xiang and Wey 2014).[33]

Also the prohibition of reproductive cloning, enshrined in all the most recent international acts on bioethics[34] and in many domestic legislations (Langlois 2017), has acquired – according to many scholars – the status of customary law (Ruffert and Steinecke 2011: 7), although there is still much controversy on the legitimacy of 'therapeutic cloning', or somatic cell nuclear transfer. In the case of reproductive cloning, over 60 countries worldwide have adopted a legislation on reproductive cloning and no country has ever legislated in order to allow this practice.

The fact that the principle of informed consent and the prohibition of reproductive human cloning have reached the status of customary rule of international law testifies to the key role played by the UNESCO declarations adopted since 1997 in the field of biomedicine in the global acceptance of fundamental principles on the protection of human rights in the sector of medicine.

Conclusion

Soft law plays a pivotal role in regulating several branches of international law, like environment, sea, space, the governance of emerging technologies such as gene editing and cyber security, and international biolaw, in the light of the 'practical' possibility it provides to reach a consensus on moral sensitive topics (since it is not legally binding). Although soft law cannot be included within the sources of international law, since it does not have

---

of the Italian Constitution (right to healthcare and not to be subjected to a forced treatment). The European Court of Human Rights in several judgements on health-related issued referred to informed consent as a fundamental principle based on the European Convention on Biomedicine (i.e. *Pretty v. the United Kingdom, Glass v. the United Kingdom, Vo v. France, Lambert Vo. France*).

32 *Rabi Abdullahi v. Pfizer, Inc.*, United States Court of Appeals for the Second Circuit; text. Available at: http://hrp.law.harvard.edu/wp-content/uploads/2011/02/abdullahi-v-pfizer-slip-op.pdf.

33 In Chinese law, the duty of a physician to request prior informed consent of a patient undergoing a medical intervention is established at Article 26, Para. 2, of the "Law of the People's Republic of China on Medical Practitioners" (1998). This principle has been subsequently recalled in guidelines and codes of conduct adopted by the China Food and Drug Administration (CFDA) and the Ministry of Health (i.e. "Ethical Review Methods for Biomedical Research Involving Humans" and "Guiding Principles for Ethical Review of Drug Clinical Trials".

34 See for instance, the UN Declaration on Human Cloning, Resolution of the World Health Organization on Cloning in Human Reproduction, Additional Protocol to the Convention on Human Rights and Biomedicine on the Prohibition of Human Cloning, Article 13 of the EU Charter on Fundamental Rights, Article 11 of the UNESCO Declaration on Human Genome and Human Rights.

any legal force, it is at the bottom of the process of formation of binding (written or non-written) rules.

In this regard, some scholars predicted the UDBHR would have been the precursor to a universal treaty on bioethics by UNESCO that would have gathered the core principles established in the three UNESCO declarations devoted to bioethics (Nys 2006: 5). However, although Nys's view did not yet come true, the evaluation on the impact of the UDBHR more than 10 years after its adoption is positive. Indeed, it constitutes – at the same 'moral' level of the Declaration of Helsinki – a foundational piece in the construction of international biomedical law, a new branch of international law that can be identified by the general recognition of human dignity as the key moral principle and of the link between bioethics and human rights (Andorno 2013b: 16).

The UDBHR can be placed within the series of international acts filling vacuums, whose provisions have not a full but at least some legal relevance. In fact, it is widely accepted that an IGO's declaration that establishes provisions in an unregulated domain of international law – in the absence of other contrary legal provisions – can set "minimum standards" widely accepted and recognized as having a legal effect (Schermers, Blokker, 2011: 788).

The UDBHR, in particular, had the merit to codify core bioethical principles that are nowadays widely accepted (the balance between benefit and harm in medical research, autonomy and individual responsibility as the basis of each medical intervention, respect for privacy and cultural diversity and pluralism) or have acquired the status of international customary law (i.e. the principle of informed consent before a medical intervention). If its impact on the European level can be considered as irrelevant, given the existing legal framework, like the Biomedicine Convention and its four additional Protocols that already established many principles of the UDBHR, or the Nizza Charter, it was a blueprint for the regulation of bioethics in developing countries. Indeed, in those countries laws and/ or regulations on bioethics have been enacted only in the last few years after the adoption of the Declaration. In this framework, the Declaration fulfilled its function of providing an adequate "framework of principles and procedures to guide states in the formulation of their legislation, policies and other instruments in the field of bioethics" (Article 2*a*).

In particular, its real innovation lies in the recognition of collective rights in the field of bioethics, in addition to the traditional individual rights related to personal autonomy. Principles such as technology transfer, benefit sharing, the need to protect the biosphere, are related to a new 'communitarian' dimension of bioethics.

However, on the other side of coin, we must point out that the excessive proliferation of soft law instruments in the field of bioethics and in related sectors (environment, biotechnology, human rights, food and agriculture, cyber security, animal welfare), denotes the lack of will by the States to limit their sovereignty in these sensitive domains that they still do consider as belonging to their domestic jurisdiction. In fact, as underlined by some scholars, the world community is witnessing a progressive detachment between the solemn enunciations of IGOs and the normative choices of the States (Campiglio 2010: 634). This trend confirms the concerns of some scholars on the stagnation in international lawmaking, previously discussed, and denotes the beginning of a new era in the lawmaking process based on the increasing recourse to informal deals to regulate the relations amongst States.

In addition, the UDBHR, like other declarations and guidelines, is weakened by the absence of enforcement procedures and penalties in case of breach. In fact, from the perspective of the victims of violations of the rights established in the UDBHR, there is a lack of a procedure that allows to litigate at courts these new rights in the field of biomedicine. In this regard, the Declaration only plays an indirect role of reference or guidance to domestic courts or international tribunals facing individual complaints for human rights violations, but it lacks direct legal authority and legal strength.

In general terms, the UDBHR inscribes itself in the current tendency of international law (from a law of coexistence to a law of positive cooperation), which amplified its sectors of intervention (Friedmann 1964). Within this tendency, States rely ever greater upon the instrument of soft law agreements; this is particularly true, as previously highlighted, for the sector bioethics and life sciences (Friedrich 2013).

As a matter of fact, soft law agreements provide at present the only realistic means of dealing with bioethical issues at a global level, and are in line with the current tendency of proliferation of informal international lawmaking, highly described and critically assessed in the literature (Pauwelyn 2012: 13).

References

Andorno, Roberto 2007a, In Defence of the Universal Declaration on Bioethics and Human Rights, *Journal of Medical Ethics*, 33: 150–154.

Andorno, Roberto 2007b, *The Invaluable Role of Soft Law in the Development of Universal Norms in Bioethics*, <www.unesco.de/wissenschaft/bis-2009/invaluable-role-of-soft-law.html>.

Andorno, Roberto 2009, Human Dignity and Human Rights as a Common Ground for a Global Bioethics, *Journal of Medicine and Philosophy*, 34: 223–240.

Andorno, Roberto 2013a, Global Bioethics at UNESCO: In Defence of the Universal Declaration on Bioethics and Human Rights, in *Health and Human Rights in a Changing World* (pp. 77–86). Grodin, Michael A., Tarantola, Daniel, Annas, George J., Gruskin, Sofia, eds., London: Routledge.

Andorno, Roberto 2013b, *Principles of International Biolaw: Seeking Common Ground at the Intersection of Bioethics and Human Rights*, Brussells: Bruylant.

Annas, George J. 1992, The Changing Landscape of Human Experimentation: Nuremberg, Helsinki and Beyond, *Health Matrix*, 2: 119–140.

Annas, George J. 2011, *Worst Case Bioethics: Death, Disaster and Public Health*, Oxford: Oxford University Press.

Arsanjani, Mahnoush H. 2006, Negotiating the UN Declaration on Human Cloning, *The American Journal of International Law*, 1: 164–179.

Asamoah, Obed Y. 1966, *The Legal Significance of the Declarations of the General Assembly of the United Nations*, Martinus Nijhoff: The Hague.

Ashcroft, Richard E. 2010, Could Human Rights Supersede Bioethics?, *Human Rights Law Review*, 10: 639–660.

Baker, Robert 1998, A Theory of International Bioethics: The Negotiable and the Non-negotiable, *Kennedy Institute of Ethics Journal*, 8: 233–273.

Barelli, Mauro 2009, The Role of Soft Law in the International Legal System: The Case of the United Nations Declaration on the Rights of Indigenous People, *International and Comparative Law Quarterly*, 58: 958–983.

Benatar, David 2005, The Trouble with Universal Declarations, *Developing Worlds Bioethics*, 5: 220–224.

Birnbacher, Dieter 2001, La Convenzione europea sulla bioetica in Germania, *Bioetica*, 9: 461–478.

Birnie, Patricia, Boyle, Alan, Redgwell, Catherine 2009, *International Law & the Environment*, Oxford: Oxford University Press.

Boczek, A. Boleslaw 2005, *International Law: A Dictionary*, Lanham, Maryla Ashgate Publishing Ltd nd, Toronto, Oxford: The Scarecrow Press Inc.

Bodansky, Daniel 2016, The Legal Character of the Paris Agreement, *Review of European, Comparative & International Environmental Law*, 25(2): 142–150.

Boussard, Hélène 2007, The Normative Spectrum of an Ethically-Inspired Legal Instrument: The 2005 Universal Declaration on Bioethics and Human Rights, in *Biotechnologies and International Human Rights* (pp. 97–128) Francioni, Francesco, ed., Oxford and Portland, Oregon: Hart.

Campiglio, Cristina, 2010, Internazionalizzazione delle fonti, in *Trattato di biodiritto, volume I* (pp. 609–635) Rodotà, Stefano, Tallacchini, Mariachiara, eds., Torino: Giappichelli.

Caporale, Cinzia 2015, Bioetica. Un futuro plurale?, *Nuova antologia*, 615: 219–223.

Cárdenas Castañeda, Fabián Augusto 2013, A Call for Rethinking the Sources of International Law: Soft Law and the Other Side of the Coin, *Anuario Mexicano de Derecho Internacional*, 13: 355–403.

Dows, George W., Rocke, David M., Barsoon, P.N. 1996, Good News about Compliance Good News About Cooperation?, *International Organization*, 50: 379–406.

Dupuy, Pierre-Marie 1991, Soft Law and the International Law of the Environment, *Michigan Journal of International*, 12: 420–435.

Dupuy, Pierre-Marie, 2007, State Responsibility for Violations of Basic Principles of Bioethics, in *Biotechnologies and International Human Rights* (pp. 33–42) Francioni, Francesco, ed., Oxford and Portland, Oregon: Hart.

Faunce, Thomas 2005, Will International Human Rights Subsume Medical Ethics? Intersections in the UNESCO Universal Bioethics Declaration, *Journal of Medical Ethics*, 31: 173–178.

Faunce, Thomas, Nasu, Hitoshi 2009, Normative Foundations of Technology Transfer and Transnational Benefit Principles in the UNESCO Universal Declaration on Bioethics and Human Rights, *Journal of Philosophy and Medicine*, 34: 296–321.

Fidler, David P. 2001, Geographical Morality Revisited: International Relations, International Law, and the Controversy Over Placebo-Controlled HIV Clinical Trials in Developing Countries, *Harvard International Law Journal*, 42: 299–354.

Fitzmaurice, Malgosia, Redgwell, Catherine 2009, Environmental Non-compliance Procedures and International Law, *Netherlands Yearbook of International Law* 31: 33–65.

Friedmann, Wolfgang 1963, The Uses of "General Principles" in the Development of International Law, *American Journal of International Law*, 57: 279–299.

Friedmann, Wolfgang 1964, *The Changing Structure of International Law*, New York: Columbia University Press.

Friedrich, Jürgen 2013, *International Environmental "Soft Law": The Functions and Limits of Nonbinding Instruments in International Environmental Governance and Law*, Heidelberg, Dordrecht, New York, London: Springer.

Goldsmith, Jack L., Posner, Erik A. 2007, *The Limits of International Law*, Oxford: Oxford University Press.

Gross, Leo 1965, Problems of International Adjudication and Compliance with International Law; Some Simple Solutions, *American Journal of International Law*, 59: 48–59.

Guzman, Andrew T. 2002, A Compliance Based Theory of International Law, *California Law Review*, 90: 1826–1888

Guzman, Andrew T. 2006, Reputation and International Law, *Georgia Journal of International and Comparative Law*, 34: 379–391.

Heriksen, Anders 2017, *International Law*, Cambridge: Cambridge University Press.

Jennings, Robert, Watts, Arthur 1992. *Oppenheim's International Law, Volume I Peace Introduction and Part 1*, Longman: Essex.

Idrissi, Nouszha Guessous 2009, Follow-Up Action by UNESCO, in *The UNESCO Universal Declaration on Bioethics and Human Rights. Background, Principles and Application* (pp. 317–326), ten Have, Henk, Stanton Jean, Michèle, eds. Paris: UNESCO Publishing.

Kirby, Michael 2009, Human Rights and Bioethics: The Universal Declaration of Human Rights and UNESCO Universal Declaration of Bioethics and Human Rights, *Journal of Contemporary Health Law & Policy*, 25: 309–331.

Klabbers, Ian 1995, The Redundancy of Soft Law, *Nordic Journal of International Law*, 65: 167.

Köch, Herbert Franz, Fischer, Peter 1997, *Das Recht der International Organisationen*, Linde Verlag: Wien.

Krasner, Stephen D. 2002, Realist Views of International Law, *Proceedings of the Annual Meeting (American Society of International Law)*, 96: 265–268.

Kratochwil, Friedrich V. 1989, *Rules, Norms, and Decisions: On the Conditions of Practical and Legal Reasoning in International Relations and Domestic Affairs*, Cambridge: Cambridge University Press.

Langlois, Adele 2013, *Negotiating Bioethics: The Governance of UNESCO's Bioethics Programme*, London and New York: Routledge.

Langlois, Adele 2017, The Global Governance of Human Cloning: The Case of UNESCO, *Palgrave Communications*, 3: 1–8.

Macpherson, Cheryl Cox 2007, Global Bioethics: Did the Universal Declaration on Bioethics and Human Rights Miss the Boat?, *Journal of Medical Ethics*, 33: 588–590.

Magnus, Richard 2016, The Universality of the UNESCO Universal Declaration on Bioethics and Human Rights, in *Global Bioethics: The Impact of the UNESCO International Bioethics Committee* (pp. 29–42) Bagheri Alireza, Moreno, Jonathan D., Semplici, Stefano, eds., Cham, Heidelberg, New York, Dordrecht, London: Springer International Publishing.

Marchant, Gary E., Allenby, Brad 2017, Soft Law: New Tools for Governing Emerging Technologies, *Bulletin of the American Scientists*, 73: 108–114.

Marchisio, Sergio 2012, *L'ONU: Il diritto delle Nazioni Unite*, Bologna: Il Mulino.

Mason, Kenyon, Laurie, Graham 2011, *Mason & McCall Smith's Law and Medical Ethics*, 8th edition, Oxford: Oxford University Press.

Nys, Herman 2006, Towards an International Treaty on Human Rights and Biomedicine? Some Reflections Inspired by UNESCO's Universal Declaration on Bioethics and Human Rights, *European Journal of Health Law*, 13: 5–8.

Pauwelyn, Joost 2012, Informal International Lawmaking: Framing the Concept and Research Questions, in *Informal International Lawmaking* (pp. 13–34) Pauwelyn, Joost, Wessel, Ramses A., Wouters Jan, eds., Oxford: Oxford University Press.

Pauwelyn, Joost, Wessel, Ramses A., Wouters, Jan 2014, When Structures Become Shackles: Stagnation and Dynamics in International Lawmaking, *European Journal of International Law*, 25(3): 733–763.

Pavone, Ilja Richard 2008, Genetica e Diritti dell'Uomo, *Enciclopedia Giuridica Treccani*, 1: 1–6.

Pavone, Ilja Richard 2009, *La Convenzione europea sulla biomedicina*, Milano: Giuffré.

Plomer, Aurora 2005, *The Law and Ethics of Medical Research: International Bioethics and Human Rights*, London: Cavendish Publishing Limited.

Posner, Erik 2014, *The Twilight of Human Rights*, Oxford: Oxford University Press.

Raustiala, Kal 2005, Form and Substance in International Agreements, *The American Journal of International Law*, 99: 581–614.

Rheeder, Riaan 2014, Article 6 of the UNESCO Universal Declaration of Bioethics and Human Rights: A moral force in South Africa, *The South African Journal of Bioethics & Law*, 7: 51–54.

Rojas Cruz, Màrcio, de Limas Torres Oliveira, Solange, Cordón Portillo, Jorge Alberto 2010, The Universal Declaration on Bioethics and Human Rights – Contributions to the Brazilian State, *Revista Bioética*, 18: 93–107.

Ruffert, Matthias, Steinecke, Sebastian 2011, *The Global Administrative Law of Science*, Heidelberg, Dordrecht, New York, London: Springer.

Shaw, Malcolm N. 2014, *International Law*, Cambridge: Cambridge University Press.

Schermers, Henry G., Blokker, Niels M. 2011, *International Institutional Law. Unity within Diversity*, Leiden and Boston: Martinus Nijhoff Publishers.

Shelton, Dina 2003, *Commitment and Compliance: The Role of Non-Binding Norms in the International Legal System*, Oxford: Oxford University Press.

Shelton, Dina 2014, International Law and "Relative Normativity", in *International Law* (pp. 141–170). Evans, Malcolm, ed., Oxford: Oxford University Press.

Sloan, Blaine 1948, The Binding Force of a Recommendation of the General Assembly of the United Nations, *British Yearbook of International Law*, 25: 1–33.

Smith, George Patrick 2012, *Law and Bioethics: Intersections Along the Mortal Coil*, Routledge: London and New York.

Sohn, Louis B. 1978, The Shaping of International Law, *Georgia Journal of International and Comparative Law*, 8: 1–25.

ten Have, Henk 2011, Foundationalism and Principles, in *The SAGE Handbook of Healthcare Ethics* (pp. 20–30) Chadwick, Ruth, ten Have, Henk, Meslin, Eric M., eds., Los Angeles, London, New Delhi, Singapore, Washington DC: SAGE.

ten Have, Henk 2015, Globalizing Bioethics Through, Beyond and Despite Governments, in *Global Bioethics: The Impact of the UNESCO International Bioethics Committee* (pp. 1–12) Bagheri, A., Moreno, J.D., Semplici, S., eds., Heidelberg, Dordrecht, New York, London: Springer.

Thürer, Daniel 2009, Soft Law – Norms in the Twilight between Law and Politics, in *International law as Progress and Prospect* (pp. 159–178) Thürer, Daniel, ed., Zurich/ St. Gallen: Dike Verlag Ag.

Thürer, Daniel 2012, *Soft Law: Max Planck Encyclopedia of Public International Law*, Vol. IX: 269-278.

UNESCO 1997, Universal Declaration on the Human Genome and Human Rights, Paris: UNESCO.

UNESCO 2003, International Declaration on Human Genetic Data, Paris: UNESCO.

UNESCO 2005, Universal Declaration on Bioethics and Human Rights, Paris: UNESCO.

Van Hoof, Godefridus 1983, *Rethinking the Sources of International Law*, Deventer, Boston: Kluwer Law and Taxation Publishers.

Williams, John R. 2005, UNESCO's Proposed Declaration on Bioethics and Human Rights – a Bland Compromise, *Developing World Bioethics*, 5: 210–215.

Xiaomei, Zhai 2009, Informed Consent in the Non-Western Cultural Context and the Implementation of Universal Declaration of Bioethics and Human Rights, *Asian Bioethics Review*, 1: 5–16.

Xiang, Yu, Wei, Li 2014. Informed Consent and Ethical Review in Chinese Human Experimentation: Reflections on the "Golden Rice Event", *Biotechnology Law Report*, 33: 155–160.

Yusuf, Abdulqawi A. 2006, The UNESCO Declaration on Bioethics: Emerging Principles and Standards of an "International Biolaw"?, in *Bioetica e biotecnologie nel diritto internazionale e comunitario* (pp. 129–139) Boschiero, Nerina, eds., Torino: Giappichelli.

IX The UNESCO Universal Declaration on Bioethics and Human Rights and the normative transition from Corporatocene to Sustainocene

Thomas Alured Faunce

Introduction

The United Nations Scientific, Education and Cultural Organization (UNESCO) *Universal Declaration on Bioethics and Human Rights* (UDBHR) involves a complex melding of three normative systems and one pseudo-normative system. Three of these normative systems, bio-ethics, international human rights and domestic law, are reasonable extrapolations of natural law social contract thought-experiments such as that influentially expressed by Rawls in his *Theory of Justice* drawing upon the work in this area of Locke, Rousseau and Kant.[1] As Rawls expressed it, the hypothetical original social contract was not one to enter a particular society or to set up a particular form of government. Rather, the guiding idea was that those establishing the prototype for civil society would decide upon principles that could be applied equally to lead to the flourishing of all. Rawls settled on two basic principles in this context. The first required equality in the assignment of basic rights and duties. The second required that social and economic inequalities, such as wealth and authority, are only just if they result in compensating benefits for everyone including the least advantaged members of society.[2] The fourth normative tradition relevant in this context involves trade and investment rights available only to multinational corporations, to claim damages from states before arbitrators. This is termed a 'pseudo-normative' tradition because the existence of a set of enforceable rules privileging artificial persons who do not vote, pay little if any tax, but start wars or take over health and educational institutions with the primary aim of maximising profit (rather than acting according to principles applicable to all) is fundamentally incompatible with the basic principles of social contract theory as developed by Locke, Rousseau, Kant and Rawls.

The hypothesis tested here is that the UDBHR is central to an emerging normative process whereby international human rights law and bioethics can be used to calibrate not only domestic judicial and statutory law, but the claims of corporations threatening or involved in trade and investment arbitral proceedings. This new normative process, it will be argued, will assist in the transition out of a governance era privileging the rights of multinational corporations (the Corporatocene) into one respecting the interests and sustainability of all life on earth (the Sustainocene).

This is a controversial thesis for several different reasons. First, as will be seen, it is difficult to determine the extent to which the UDBHR has emerged out of a solid theoretical foundation in academic bioethics or international human rights and, if it has, whether that conceptual

1 J. Rawls, *Theory of Justice*, Oxford University Press, Oxford 1971.
2 Ibid., 14–15.

backbone should be categorised as coherent with the positivist systems of law which defines law by reference to a constitutional rule of recognition and the capacity to distinguish it from private morals. Second, the sovereign importance of any normative system utilising constitutional provisions, legislation and judicial decisions is now significantly challenged by the power of multinational corporations to claim damages before arbitral panels using dispute settlement mechanisms in trade and investment agreements such as the *North American Free Trade Agreement* (NAFTA),[3] the *Energy Charter Treaty* 1994[4] and the *Trans Pacific Partnership Agreement*.[5]

Background to the UDBHR

In 2003, UNESCO released a report it had commissioned from its international bioethics committee on the possibility of elaborating a Universal Declaration or Convention on Bioethics.[6] This was the outcome of work that had begun with a resolution of the UNESCO General Conference at its 31st Session, calling on its Director-General to submit "the technical and legal studies undertaken regarding the possibility of elaborating universal norms on bioethics".[7]

To assist in the production of a draft text, a team of eminent international scholars was appointed under the chairmanship of Justice Michael Kirby of the High Court of Australia. To further assist Justice Kirby, on 18–19 November 2004 a meeting of both bioethics and international human rights experts was convened by the author at Manning Clark House (MCH) in Canberra, Australia, to discuss a draft text.[8] The MCH experts meeting recommended that substantial consideration be given as to whether the UDBHR was to be the precursor to a UNESCO *International Convention on Bioethics and Human Rights* involving binding norms under public international law for those nations who signed and ratified. If so, they argued, then there would be major ramifications for the normative frameworks of both bioethics and public international law, including their application to multinational corporate actors. The development of the UDBHR was prescient given that a large part of international human rights law was now concerned with health-related matters that strongly overlapped with bioethical concerns and principles.[9]

The resultant *Universal Declaration on Bioethics and Human Rights* (UDBHR) is what is known as a 'non-binding' declaration under public international law, insofar as that discipline

---

3 www.naftanow.org, last accessed 02-08-2017.

4 www.encharter.org, last accessed 02-08-2017.

5 http://dfat.gov.au/trade/agreements/tpp/pages/trans-pacific-partnership-agreement-tpp.aspx, last accessed 02-08-2017.

6 UNESCO, *Report of the IBC on the Possibility of Elaborating a Universal Instrument on Bioethics*, SHS/EST/02/CIB-9/5 (Rev. 3), Paris, 13 June 2013.

7 UNESCO, *31C/Resolution: Bioethics Programme: Priorities and Perspectives*, UNESCO General Conference, 31st Session, Paris, 2002. UNESCO, 'Report of the IBC on the Possibility of Elaborating a Universal Instrument on Bioethics,' SHS/EST/02/CIB-9/5 (rev 3), 2003. UNESCO, 'Elaboration of the Declaration on Universal Norms on Bioethics,' Third Outline of a Text SHS/EST/04/CIB-Gred-2/4 rev.2, Paris, 2004.

8 MECHM Manning Clark House Meeting of Bioethics, Health Law and International Human Rights Experts Consensus Statement on the UNESCO Universal Bioethics Declaration. 18–19 November 2004.

9 United Nations, 1966, *International Covenant on Civil and Political Rights*. Adopted 16 Dec 1966, entry into force 23 March 1976. GA Res 2200A (XXI). UN GAOR supp (no 16) 52. UN doc A/6316. *UNTS*; 999: 17. United Nations, 1966, *International Covenant on Economic, Cultural and Social Rights*. Adopted 16 Dec 1966, entry into force 3 Jan 1976. GA Res 2200A(XXI). UN Doc A/6316, *UNTS*; 993: 3 United Nations. (1945). *United Nations Statute of the ICJ* 1945 *UNTS*; 1: xvi. United Nations. (1948). *Universal Declaration of Human Rights*. Adopted 10 Dec 1948. GA Res 217A (III). UN doc A/810 (1948) 71.

is defined by Article 38 of the *Statute of the International Court of Justice*.[10] Another way of describing the UDBHR is to call it 'soft law,' a controversial term generally referring to a loosely defined category of putative norms for which states commit to merely having a legitimate interest in mutual compliance, rather than any formal undertaking of enforceable obligations (whose role in bioethics has been described by Pavone in the previous chapter of this book).[11,12,13] Thus 'soft-law' could apply to a range of quasi-legal international norms from constructive ambiguities (such as corporate reward for 'innovation,' or encouragement to regulatory 'transparency' in multilateral and bilateral trade and investment agreements), to guidelines and standards of measurement by expert panels or committees of intergovernmental organisations or peak non-governmental organisations.[14] The decision to create an ostensibly hybrid bioethics–human rights text such as the UDBHR may be viewed, indeed, as a complex geo-political trade-off amongst those elected and corporate oligarchies controlling nations in the decentralised, non-hierarchical international governance system.[15]

Article 1 of the UDBDR indicates that the principles of that text are not just addressed to states (as would be expected of a public international law document) but also to individuals, communities and corporations. It reads:

Article 1 – scope

1. This Declaration addresses ethical issues related to medicine, life sciences and associated technologies as applied to human beings, taking into account their social, legal and environmental dimensions.
2. This Declaration is addressed to states. As appropriate and relevant, it also provides guidance to decisions or practices of individuals, groups, communities, institutions and *corporations*, public and private.

[emphasis added]

Article 2 indicates that the UDBHR has a variety of aims, which include legal, ethical and political objectives. Articles 2 (c) and (d) also refer to promotion of respect for human dignity distinctly from protection of human rights.

Article 2 – aims

The aims of this Declaration are:

(a) to provide a universal framework of principles and procedures to guide states in the formulation of their legislation, policies or other instruments in the field of bioethics;

10 B. Simma, P. Alston, The Sources of Human Rights Law: Custom, Jus Cogens and General Principles, *Australian Yearbook of International Law*, 1992, vol. 12, pp. 82–102.
11 J. Klabbers, The Redundancy of Soft Law, *Nordic Journal of International Law*, 1996, vol. 67, pp. 167–178.
12 J. Mann, Dignity and Health: The UDHR's Revolutionary First Article, *Health and Human Rights*, 1998, vol. 3, no. 2, pp. 31–38.
13 J. Raz, The Nature of Rights, *Mind*, 1984, vol. 93, pp. 194–202.
14 T. A. Faunce, Will International Human Rights Subsume Medical Ethics? Intersections in the UNESCO Universal Bioethics Declaration, *Journal of Medical Ethics*, 2005b, vol. 31, pp. 173–178.
15 K. Raustiala, Form and Substance in International Agreements, *The American Journal of International Law*, 2005, vol. 99, no. 3, pp. 581–614.

(b) to guide the actions of individuals, groups, communities, institutions and *corporations*, public and private;

(c) to promote respect for human dignity and protect human rights, by ensuring respect for the life of human beings, and fundamental freedoms, consistent with international human rights law;

(d) to recognise *the importance of freedom of scientific research and the benefits derived from scientific and technological developments*, while stressing the need for such research and developments to occur within the framework of ethical principles set out in this Declaration and to respect human dignity, human rights and fundamental freedoms;

(e) to foster multidisciplinary and pluralistic dialogue about bioethical issues between all stakeholders and within society as a whole;

(f) to *promote equitable access to medical, scientific and technological developments* as well as the greatest possible flow and the rapid *sharing of knowledge* concerning those developments and the *sharing of benefits*, with particular attention to the needs of developing countries;

(g) to safeguard and promote the interests of the present and future generations;

(h) to underline the *importance of biodiversity and its conservation* as a common concern of humankind.

[emphasis added]

The UDBHR, as we shall see, contains socially important principles supporting equity in technology and knowledge transfer as well as transnational benefit, particularly in Articles 14, 15 and 21. The normative foundations of such principles or norms and whether they can legitimately be called such under public international law or bioethics are controversial topics. Some theoretical bioethical justifications for UDBHR technology transfer statements could be consequentialist, based on potential adverse outcomes for national security of an inadequate response. Others could be deontological – linked to the foundational need in a nominally liberal society for institutions to respect basic elements of human capacity and functioning and to inspire efforts for the public good. Some justifications could be virtue-oriented: supporting justice, fairness, respect for human dignity as character traits that should manifest in a well-ordered society as they do in people who apply culturally valued principles consistently in the face of obstacles. Legal positivists would generally consider the UDBHR as not formed in connection with a legal rule of recognition thus emphasising pursuit of social virtues such as certainty and consistency rather than justice and equity. Nonetheless even they would consider that the UDBHR principles could influence the creation of legal norms.

As mentioned earlier, the thesis being advanced here is that UDBHR principles directed at governments and at corporations encouraging equity technology transfer and transnational benefit will have an important role in facilitating the transition from Corporatocene to Sustainocene. It is now necessary to consider those latter terms in more detail.

Corporatocene to Sustainocene transition

The term 'Holocene' ("recent whole") was attached to the post-glacial geological epoch by the International Geological Congress in Bologna in 1885. It is defined as beginning 10,000 years ago. From that time until about 1800 CE, humanity's activities barely changed the natural systems of this world. Since 1800 with the onset of the Industrial Revolution, the development of the capacity to fix atmospheric nitrogen as a fertilizer, improved sanitation health care and transport, increasing human population and its impact, have dramatically enhanced our capacity to extinguish other species, burn photosynthesis fuels archived over millions of years (in the form of coal, oil and natural gas) thereby increasing greenhouse gas concentration of CO_2 in the atmosphere, as well as destroy and convert land ecosystems to cities of bitumen and asphalt. In the 1920s V. I. Vernadsky, P. Teilhard de Chardin and E. Le Roy devised the term 'noösphere' (the world of thought) to emphasise the growing role played by humankind's brainpower and technological talents in shaping its own future and environment.[16]

In what seems to be an extension of the noösphere idea, it has been argued that human activity has pushed this planet from the Holocene into what has been termed the 'Anthropocene' period, a term coined by Crutzen in 2002.[17] 'Anthropocene' refers to an epoch when human interferences with earth systems (particularly in the form of influences on land use and land cover, coastal and maritime ecosystems, atmospheric composition, riverine flow, nitrogen, carbon and phosphorus cycles, physical climate, food chains, biological diversity and natural resources) have become so pervasive and profound that they are not only becoming the main drivers of natural processes on earth, but are threatening their capacity to sustain life.[18] Salutary facts driving academic and policy interest in moving from the Anthropocene to a different type of human-controlled epoch are not only the anthropogenic greenhouse-gas-driven increase in severe weather events, but the projected increase of global human population to around 10 billion by 2050 with associated energy consumption rising from ≈400EJ/yr to over 500EJ/yr beyond the capacity of existing fossil-fuel-based power generation.[19] The research underpinning the push to develop an environmentally better energy and climate policy also emerged strongly from influential commentaries such as the Intergovernmental Panel on Climate Change[20] and the Stern Report.[21] In the Paris Accord of 2015 (into force since 2016) governments of most of the world's nations agreed to 1) a long-term goal of keeping the increase in global average temperature to well below 2°C above pre-industrial levels; 2) to aim to limit the increase to 1.5°C, since this would significantly reduce risks and the impacts of climate change; 3) on the need for global emissions to peak as soon as possible, recognising that this will take longer for developing countries; 4) to undertake rapid reductions thereafter in accordance with the best available science.[22]

But is 'Anthropocene' an accurate term? Five features of the Anthropocene epoch are alleged to dominate its policy debates: population, poverty, preparation for war, profits and pollution.[23] Of these, every one except the first, overpopulation, is a direct outcome of the increasing socio-political influence and desire to maximise shareholder profits and executive

16 P. J. Crutzen, E. Stoermer, *Global Change Newsletter*, 2000, vol. 41, p. 17.

17 P. J. Crutzen, Geology of Mankind, *Nature*, 2002, vol. 415, p. 23.

18 W. Steffen, P. J. Crutzen, J. R. McNeill, The Anthropocene: Are Humans Now Overwhelming the Great Forces of Nature, *AMBIO: A Journal of the Human Environment*, 2007, vol. 36, no. 8, p. 614.

19 H. H. Rogner, United Nations Development World Energy Assessment, *United Nations, Geneva*, 2004, vol. 5, p. 162.

20 R. K Pachauri, A. Reisinger eds., *Report of the Intergovernmental Panel on Climate Change*, Geneva, Switzerland, 2007.

21 N. Stern, *The Economics of Climate Change: The Stern Review: Cabinet Office HM – Treasury*, Cambridge University Press, Cambridge, UK, 2007.

22 European Commission, *Climate Action. Paris Agreement*, 2015. Available at: https://ec.europa.eu/clima/policies/international/negotiations/paris_en (accessed July 2017). The Paris Agreement is now challenged by the decision of President Trump to withdraw from the Agreement (https://www.whitehouse.gov/blog/2017/06/01/president-donald-j-trump-announces-us-withdrawal-paris-climate-accord; last visited July 2017).

23 B. Furnass, *From Anthropocene to Sustainocene*. Public Lecture. Australian National University, 21 March 2012.

remuneration of multinational corporations. It is not the average citizen who is responsible for an oversupply of food that is dumped in one part of the globe while in another part people starve to death. It is likewise not the responsibility of average people but of profit-seeking multinational armaments manufacturers that many wars break out and cause such great loss of life. Pollution on a grand scale is more than anything else a problem created by oil and coal and mining companies, plastic manufacturing companies, the global agrifood business. Looked at critically it is more appropriate to term the Anthropocene the 'Corporatocene'.

The dominant political and social actor in the Corporatocene is the multinational corporation. Such artificial human entities significantly erode the sovereignty of the state. They do this by large donations to political parties who in turn ensure a process of turning over public assets to corporate hands (privatisation), preventing the establishment of new public assets (through requirements under trade and investment agreements to compensate corporate actors for loss of investment), use of judges, police and military to enforce patents and create wars to maintain profit, facilitate the transfer of money by the wealthy to offshore tax havens as well as by inhibiting the development of governance arrangements or new technologies that would hamper this process.

In the early 1990s, civil society prevented the creation of a supranational investment protection agreement (the *Multilateral Agreement on Investment* or MIA) that would have allowed the global implementation of principles that allowed supranational corporations to sue (before small panels of commercial arbitration lawyers with little understanding of or desire to apply international public law) other nations who have imposed governance requirements (even when in the public health and environmental interest based on good scientific evidence) if their commercial interests are thereby impeded. Such investor-state provisions surfaced again in the 1994 *North American Free Trade Agreement* (NAFTA) between the United States (US), Canada and Mexico.[24] They are now part of over 2,000 bilateral investment treaties (BITs).[25] They are probably the main reason the US Trade Representative sought to create the *Trans Pacific Partnership Agreement* and will lobbied for as a major inclusion in the Trans-Atlantic Trade and Investment Agreement. They grant investors covered by them a right to initiate dispute settlement proceedings (before a panel of trade lawyers known as commercial arbiters) for damages against foreign governments in their own right.[26]

The conflicted individuals officiating on such arbitral proceedings view such investment agreements as private contracts, are paid by the parties and do not necessarily take account of domestic public health and environment protections – creating a pro-investor jurisprudence. Investor-state challenges have occurred in relation to a broad spectrum of public health and the environment legislation and policies. Statutes on water protection, waste disposal and waste treatment as well as universal health care and access to affordable medicines have been challenged by supranational corporations seeking damages under investor-state mechanisms on the basis that such legislation actually or potentially erodes the investments of multinational corporations in a country. Investor-state provisions have been criticised as

24 AFTINET TPP submission to Minister for Trade viewed March 2010 www.dfat.gov.au/trade/fta/tpp/subs/tpp_sub_aftinet_081103.pdf. Ranald and Southalan, The Australia–US Free Trade Agreement: Trading Australia Away? *Australia Fair Trade and Investment Network*. Available at: www.bilaterals.org/IMG/pdf/ranald.pdf, last accessed 02-08-2017.
25 S. Ganguly, The Investor-State Dispute Mechanism (ISDM) and a Sovereign's Power to Protect Public Health, *Columbia Journal of Transnational Law*, 1999, vol. 38, p. 113.
26 UNCTAD, *Dispute Settlement: Investor–State*. United Nations Publication 2003. Available at: www.unctad.org/en/docs/iteiit30_en.pdf, viewed April 2010.

allowing foreign investors leverage to undermine government legislation promoting, for example, sustainable development, environmental protection and public health policy.[27] Investor-state dispute settlement claims have challenged attempts by nation states to regulate against chemicals proven to cause developmental disability,[28] neurotoxins,[29] hazardous lawn pesticides[30] and carcinogenic gasoline additives.[31] The mechanism has also been used by foreign corporations to attempt to overturn legislation on water security,[32] waste disposal,[33] waste treatment[34] and a US ban on cattle with suspected bovine spongiform encephalopathy (BSE or mad cow disease).[35]

The investor-state legal mechanism sits in a twilight zone between international public law (including international human rights law) and commercial arbitration. Tobacco manufacturer Philip Morris's investor-state claim against Australia's plain packaging legislation (scientifically proven to reduce the rate of youth smokers) is an example of how such provisions cut across bioethical and international human rights principles related to the right to health. The United States registered corporations (even those who have their headquarters in other nations for tax purposes) have rarely if ever lost an investor-state dispute settlement (ISDS) claim; indeed endorsing this meta-understanding appears to be an implicit prerequisite for selection as a trade and investment arbitrator.

Supranational corporations undoubtedly could use this ISDS mechanism to claim compensation where a new technology for distributed food, fuel and fertilizer (such as global artificial photosynthesis) was subsidised by a government on the basis that its products were more environmentally friendly or safe from a public health point of view.

Humanity in the Corporatocene developed the capacity to diagnose what may be termed 'planetary illness', by robust measures such as biodiversity loss (and species extinction), atmospheric carbon dioxide levels, availability of fresh water.[36] Such tests resemble those that allowed medical science to diagnose human illness in the 19th century, a period when few effective remedies were in existence.

27 D. Esty, Bridging the Trade-Environment Divide, *Journal of Economic Perspectives*, 2001, vol. 15, no. 3, pp. 113–130.

28 NAFTA – Chapter 11, Investment Cases Filed Against the Government of Canada Cromptom (Chemtura) Corp v Government of Canada viewed April 2010. Available at www.international.gc.ca/trade-agreements-accords-commerciaux/disp-diff/crompton_archive.aspx?lang=en.

29 US Department of State. Ethyl Corp v Government of Canada. Available at www.state.gov/s/l/c3745.htm 12 April 2010.

30 US Department of State Dow AgroSciences LLC v. Government of Canada. Available at www.state.gov/s/l/c29885.htm 12 April 2010.

31 Governor of California Executive Order D-5-99, March 25 1999 cited in S. Gaines, Methanex Corp v United States, *American Journal of International Law*, July 2006, vol. 100, no 3, pp. 683–689.

32 Sun Belt Water Inc Notice to submit a claim to arbitration under Chapter 11 NAFTA. Viewed April 2010. Available at www.international.gc.ca/trade-agreements-accords-commerciaux/topics-domaines/disp-diff/sunbelt.aspx?lang=eng, viewed August 2017.

33 US Department of State. V. G. Gallo v Government of Canada. Available at www.state.gov/s/l/c29744.htm 12 April 2010.

34 Metalclad Corp v United Mexican States. Available at www.state.gov/s/l/c3752.htm 12 April 2010.

35 US Department of State Cases regarding the border closures due to BSE concerns. Available at www.state.gov/s/l/c14683.htm viewed April 2010.

36 W. Steffen, A. Persson, L. Deutsch, J. Zalasiewicz, M. Williams, K. Richardson, C. Crumley, P. Crutzen, C. Folke, L. Gordon, M. Molina, V. Ramanathan, J. Rockström, Johan, M. Scheffer, H.J. Schellnhuber, U. Svedin, 2011, The Anthropocene: From Global Change to Planetary Stewardship, *AMBIO: A Journal of the Human Environment*, 2011, vol. 40, p. 739

A terminological revision from Holocene to Corporatocene focuses public awareness and policy attention more precisely on for-profit multinational corporations, the core of the problem here for democratic governance and environmental sustainability.[37] It also focuses on the need to move to a different type of vision and system – one that reverences all life on earth and in which the major players in political power seek to consistently apply universally applicable principles. It further encourages human innovation to develop technological therapies for the global problems the earth faces. Such a vision is that of the Sustainocene.

The Sustainocene

The term 'Sustainocene' was coined by the Canberra-based Australian physician Bryan Furnass in 2012.[38] It has been described as referring to a period where governance structures and scientific endeavour coordinate to achieve the social virtues of ecological sustainability and environmental integrity as influentially propounded by eco-economists such as E. F. Schumacher (with his concept of 'small (and local) is beautiful') and Kenneth Boulding (with his idea of 'Spaceship Earth' as a closed economy requiring recycling of resources) as well as Herman Daly with his notion of 'steady state' economies drawing upon the laws of thermodynamics and the tendency of the universe to greater entropy (dispersal of energy).[39]

One area of academic research and policy development that fits well with "Sustainocene" thinking is that centred on the idea that this planet should be treated not just as a distinct living entity (James Lovelock's Gaia Hypothesis), but as a patient.[40] 'Planetary medicine' as this field is known has become a symbolic rubric focusing not just public and governmental attention on the interaction between human health, technological development and sustainability of the biosphere.[41] In this emerging discipline, characteristic features of the Corporatocene epoch such as anthropogenic climate change and environmental degradation, as well as gross societal imbalances in poverty and lack of necessary fuel, food, medicines, security and access to nature, are targeted as intrinsically global pathologies, the resolution of which requires concerted efforts to implement a wide range of not just renewable energy technologies but bioethical principles including those related to protecting the interests of future generations and preservation of biodiversity. One of the major differences between the Corporatocene and the Sustainocene may be that in the latter humanity was able to develop a planetary therapeutic: notably global artificial photosynthesis ('AP'). This is arguably one of the most significant technologies to which the UDBHR provisions will apply.

Sustainocene and global artificial photosynthesis

When we travel in aircraft across the world it is easy to see the extent to which human concrete and asphalt structures are proliferating across the face of the planet. Such structures contribute little to the ecosystems around them. They do not enrich the soil or provide

37 T.A. Faunce, *Who Owns Our Health. Medical Professionalism, Law and Leadership Beyond the Age of the Market State*, University of New South Wales Press: Sydney, 2007.

38 B. Furnass, *From Anthropocene to Sustainocene: Challenges and Opportunities*. Public Lecture. Australian National University, Sidney, 21 March 2012.

39 T. A. Faunce, *Nanotechnology for a Sustainable World: Global Artificial Photosynthesis as the Moral Culmination of Nanotechnology*, Edward Elgar, Cheltenham 2012.

40 J. E. Lovelock, *Gaia, the Practical Science of Planetary Medicine*, Gaia Books, London 1991.

41 T. McMichael, The Biosphere, Health, and "Sustainability", *Science*, 2002, vol. 297, no. 5584, p. 1093.

oxygen or absorb carbon dioxide. Yet we are almost at the point where nanotechnology and artificial photosynthesis can be engineered into such structures so they can be made to "pay their way" in an ecosystem sense.

The material preconditions for offering global artificial photosynthesis as a planetary therapeutic for the Sustainocene are strong. More solar energy strikes the earth's surface in one hour of each day than the energy used by all human activities in one year.[42] At present the average daily power consumption required to allow a citizen to flourish with a reasonable standard of living is about 125kWh/day. Much of this power is devoted to transport (~40 kWh/day), heating (~40 kWh/day) and domestic electrical appliances (~18 kWh/day), with the remainder lost in electricity conversion and distribution.[43] Global energy consumption is approximately 450 EJ/yr, much less than the solar energy potentially usable at ~1.0 kilowatts per square metre of the earth – 3.9×10^6 EJ/yr even if we take into the earth's tilt, diurnal and atmospheric influences on solar intensity.[44] The question of how best to use this solar energy remains a major contemporary policy conundrum. Photovoltaic (PV) energy systems (which put solar photons into batteries, or the electricity grid) are improving their efficiencies towards 25%, and the cost of the electricity they produce is nearing or has past grid parity in many nations. The development of "smart-grid" (allowing energy-carrying capacity to fluctuate coherently in accord with renewable source input and output) and "pumped-hydro" (using diurnal PV electricity to pump water to high reservoirs so it can be run down through turbines at night) will assist the viability of this as a national energy source. Even large solar farms, however (for example taking up 200 m² per person with 10%-efficient solar panels) could produce but ~50kWh/day per person.[45] Yet the problem has been solved by plants a billion years ago – to use it to make fuel and food locally in the same organism that captures the light, by drawing upon the resource of atmospheric carbon dioxide.

Similarly, there has been much policy interest in developing what is termed the 'hydrogen economy' in which hydrogen is used ubiquitously as a carbon-neutral energy vector (for example source of electricity via fuel cells or as a fuel itself when combined with atmospheric nitrogen to form ammonia) and source of fresh water (when combusted). Major policy documents have outlined the case for a hydrogen economy.[46,47,48,49] Significant scientific challenges here include the need to lower the cost of hydrogen fuel production to that of petrol, the difficulties in creating a sustainable and low carbon dioxide route for the mass production of hydrogen, the need to develop safe and more efficient storage (including the difficulties of compressing and cooling the hydrogen), the need to develop regulations and

42 L. Hammarström, S. Hammes-Schiffer, Artificial Photosynthesis and Solar Fuels, *Accounts of Chemical Research*, 2009, vol. 42, no. 12, p. 1859.
43 D. J. C. MacKay, *Sustainable Energy-Without the Hot Air*, UIT, Cambridge, 2009, p. 204.
44 A. B. Pittock, *Climate Change: The Science, Impacts and Solutions*, 2nd edition, CSIRO Publishing. Collingwood, 2009, p. 177.
45 D. J. C. MacKay, *op. cit.*, p. 41.
46 European Hydrogen and Fuel Cell Technology Platform. Available at www.hfpeurope.org/hfp/keydocs.
47 US Department of Energy: Hydrogen Posture Plan. Available at www.hydrogen.energy.gov/.
48 The Hydrogen Economy: Opportunities, Costs, Barriers and R&D Needs, National Research Council and National Academy of Engineering Report. Available at www.nap.edu/catalog/10922.html.
49 E4tech, Element Energy, Eoin Lees Energy, A Strateguc Framework for Hydrogen Energy in the UK. Available at www.berr.gov.uk/files/file26737.pdf.

safety standards at national and international levels as well as the need to develop stable incentive systems for large-scale investment in this area that will not fluctuate with oil prices.

One of the main problems at present with moving to a global hydrogen economy is the carbon-intensive energy required to produce hydrogen in large quantities by steam reformation of hydrocarbons, generally methane. Hydrogen (H_2) on a weight basis has three times the energy content of gasoline. Liquifying H_2 requires a complex and expensive process in which approximately 35% of H_2 energy is lost. Compression of H_2 similarly requires considerable external energy and a cylindrical shape.[50] This problem partially may be solved by considering the vast nitrogen resource constituting 78% of the atmosphere. Hydrogen and nitrogen can be combined to make ammonia – a valuable fuel and source of fertilizer.

Making an abundant and easily accessible form of hydrogen (ATP) by the splitting of water using energy from the sun is the second and most important aspect of solving the puzzle. The process of photosynthesis is that by which plants (i.e. cyanobacteria now preserved as stromatolite fossils in places such as Shark Bay in Western Australia and Zebra River in Namibia) have created the ecosystems of the earth. Photosynthesis (in its traditional form utilising biology) provides the fundamental origin of our oxygen, food and the majority of our present-day fuels; it has been operating on earth for 2.5 billion years.[51]

The process of doing photosynthesis is so well understood that it is a feasible scientific challenge to not only replicate it but improve upon it.[52] The capacity to store solar energy in transportable chemical bonds is the feature that makes enhanced photosynthesis so intriguing as a form of renewable energy.

Photosynthesis can be considered as a process of planetary respiration: breathing in it creates a global annual CO_2 flux[53] and on expiration an annual O_2 flux.[54] In its present nanotechnologically unenhanced form, photosynthesis globally already traps around 4,000 EJ/yr solar energy in the form of biomass.[55] The global biomass energy potential for human use from photosynthesis as it currently operates globally is approximately equal to human energy requirements (450 EJ/yr).[56,57,58]

Biologic photosynthesis is a research trial that has been successfully conducted by life on earth for billions of years. It would be sensible to consider improving upon it as a likely pathway to energy security and environmental sustainability for humanity.

At the same time as the puzzle of how to do photosynthesis most effectively began to exercise the minds of some scientists, humanity developed a revolutionary approach to making things – nanotechnology. Nanotechnology is the science of making things from

50 A. Sartbaeva, V. L. Kuznetsov, S. A. Wells, P. P. Edwards, Hydrogen Nexus in a Sustainable Energy Future, *Energy and Environmental Science*, 2008, vol. 1, pp. 79–85.

51 M. Leslie, Origins: On the Origin of Photosynthesis, *Science*, 2009, vol. 323, p. 1286.

52 R. E. Blankenship, *Molecular Mechanisms of Photosynthesis Blackwell Science*, Blackwell Science, Oxford/ Malden 2002.

53 C. Beer, M. Reichstein, E. Tomelleri, P. Ciais, M. Jung, N. Carvalhais, Terrestrial Gross Carbon Dioxide Uptake: Global Distribution and Covariation with Climate, *Science*, 2010, vol. 329, p. 834.

54 W. Hillier, T. Wydrzynski, *Cordination Chemistry Review*, 2008, vol. 252, p. 306.

55 A. Kumar, D. D. Jones, M. A. Hann, Thermochemical Biomass Gasification: A Review of the Current Status of the Technology, *Energies*, 2009, vol. 2, p. 556.

56 M. Hoogwijk, A. Faaij, R. van den Broek, G. Berndes, D. Gielen, W. Turkenburg, Exploration of the Ranges of the Global Potential of Biomass for Energy, *Biomass Bioenergy*, 2003, vol. 25, p. 119.

57 M. Parikka, Global Biomass Fuel Resources, *Biomass Bioenergy*, 2004, vol. 27, p. 613.

58 G. Fischer, L. Schrattenholzer, Global Bioenergy Potentials Through 2050, *Biomass Bioenergy*, 2001, vol. 20, p. 151.

components that are not much bigger than a few atoms, less than 100nm (a nanometer is a billionth of a metre). The chief policy interest to date with nanotechnology has been concerned with ensuring its safety.[59] Corporations (as was to be expected in the Corporatocene) have focused on making money from nanotechnology through consumer products such as lightweight, strong sporting goods (carbon fibre golf clubs and racing bikes) and odourless socks and shirts as well as packaging that preserves food as it is flown or container-shipped around the world (with nanosilver).

Experts, however, have encouraged nanotechnology researchers instead to systematically contribute to achievement of the United Nations *Millennium Development Goals* (as they then were), particularly energy storage, production and conversion, agricultural productivity enhancement, water treatment and remediation.[60] Nanotechnology could equally be prioritised to focus on achievement of the *Sustainable Development Goals*. 1) One in five people still lacks access to modern electricity, 2) 3 billion people rely on wood, coal, charcoal or animal waste for cooking and heating, 3) Energy is the dominant contributor to climate change, accounting for around 60% of total global greenhouse gas emissions and 4) Reducing the carbon intensity of energy is a key objective in long-term climate goals.[61]

Yet the case can be made that looked at from an idealistic perspective coherent with basic ethical and human rights principles, the moral culmination of nanotechnology should be global artificial photosynthesis ('GAP').[62] In simple terms ethics is a process of developing principles that can be consistently applied by all rational persons to produce virtue and mutual flourishing. In such basic ethical terms if humanity breathes oxygen, its buildings should make oxygen. Ethically, if humanity breathes out carbon dioxide, its buildings should resorb from the atmosphere that greenhouse gas. The idea of making the all human structures on the earth's surface do photosynthesis without biology is an ethical commitment at the core of the vision of a transition to a Sustainocene epoch. In this way technology operating at a billionth of a metre can improve upon and take some economic pressure off a biological system successfully operating for billions of years. The development of an economy based on practical solar fuels (whether focusing primarily on splitting water to create hydrogen, or also utilising atmospheric nitrogen to make ammonia) will be a major step in shifting the biosphere from what has been termed the 'Corporatocene' to the Sustainocene epoch. It will no doubt also become a governance battlefield as corporations earning vast profits from oil, coal and fracking gas attempt to use patent laws and ISDS proceedings to prevent governments from facilitating deployment of such new renewable energy and climate change mitigation technologies in accordance with principles of justice and fairness such as those in the UDBHR.

Many researchers in the AP field will continue for several decades to consider that genetically modifying or utilising plants can and will continue to be the best option. They will seek for example to genetically manipulate or even synthetically reproduce photosynthetic

59 T. A. Faunce, Three Proposals for Rewarding Novel Health Technologies Benefiting People Living in Poverty: A Comparative Analysis of Prize Funds, Health Impact Funds and a Cost-Effectiveness/ Competitive Tender Treaty, *Expert Opinion in Drug Safety*, 2008, vol. 7, no. 2, pp. 103–106.
60 F. Salamanca-Buentello, Nanotechnology and the Developing World, *PloS Med.*, 2005, vol. 2, p. e97.
61 United Nations, *Sustainable Development Goals Goal 7 Energy*. Available at: www.un.org/sustainable development/energy/.
62 T. A. Faunce, *op. cit.*

plants and bacteria to maximise their light capture and carbon reduction activities.[63] This is likely to remain an attractive area because scientists will be able to deliver results in short grant cycles. Yet long term the AP field will begin to shift towards non-biological nanotechnology-based AP. This is not simply because the scientific challenge of under-standing and replicating natural AP is intriguing but there are significant implications of being able to capture many more photons than natural systems, to use them more efficiently to make fuel and food and fertilizer not only from atmospheric carbon dioxide, but from atmospheric nitrogen.

One model of a Sustainocene powered by solar fuels involves bio-mimetic polymer pho-tovoltaic generators plugged into the national electricity grid to power (near large sources of seawater, CO_2, waste heat, high solar irradiation and proximity to end use facilities) large-scale hydrogen fuel and waterless agriculture, chemical feedstocks and polymers for fibre production.[64] This model has the advantage of the 'light' and 'dark' reactions being uncoupled in relation not only to energy/material flow balance, but also to the requirement to be co-located in space. Yet such a model favours power concentration in the hands of a few and in that sense is less ethical than a model of GAP which emphasises individual and community involvement in micro or local generation of fuel and food through AP products installed as a policy priority on domestic dwellings and vehicles.[65] There is a simple ethical message at the core of the Sustainocene in telling people that nanotechnology will be used to make buildings function like trees. A device that can do this and is available to cheap purchase and installation, like the mobile phone or Internet, could rapidly transform society into a place more characterised by virtues like equity and environmental sustainability.

Yet if global artificial photosynthesis is to be utilised this way, governance principles such as equitable transfer of such transformative technologies will need to be in place.[66,67] It is here that the UDBHR principles on equity in technology transfer addressed to corpora-tions could have a critical role in developing positive law domestic norms to counteract the pseudo-legal norms developing at the behest of multinational corporations out of trade and investment agreements.

Technology transfer and transnational benefit in the UDBHR

The UDBHR contains three substantial provisions on technology transfer and transnational benefit likely to assist in shaping the principles by which new renewable energy and climate

63 A. J. Ragauskas, C. K. Williams, B. H. Davison, G. Britovsek, J. Cairney, C. A. Eckert, W. J. Frederick Jr, J. P. Hallett, D. J. Leak, C. L. Liotta, J. R. Mielenz, R. Murphy, R. Templer, T. Tschaplinski, The Path Forward for Biofuels and Biomaterials, *Science*, 2006, vol. 484, p. 311.
64 R. Pace, An Integrated Artificial Photosynthesis Model, in A. Collings, C. Critchley (eds.), *Artificial Photosynthesis: From Basic Biology to Industrial Application*, Wiley-VCH Verlag, Weinheim 2005, p. 13.
65 T. A. Faunce, Ch 21: Future Perspectives on Solar Fuels, in T. Wydrzynski, W. Hillier (eds.), *Molecular Solar Fuels Book Series: Energy*. Royal Society of Chemistry, Cambridge UK 2012, pp. 506–528.
66 T. A. Faunce, W. Lubitz, A. W. Rutherford, D. MacFarlane, G. F Moore, P. Yang, D. G. Nocera, T. A. Moore, D. H. Gregory, S. Fukuzumi, K. B. Yoon, F. A. Armstrong, M. R. Wasielewski, S. Styring, Energy and Environment Policy Case for a Global Project on Artificial Photosynthesis, *Energy and Environmental Science*, 2013, vol. 6, no. 3, p. 695.
67 T. A. Faunce, S. Styring, M. R. Wasielewski, G. W. Brudvig, A. W. Rutherford, J. Messinger, A. F. Lee, C. L. Hill, H. deGroot, M. Fontecave, D. R. MacFarlane, B. Hankamer, D. G. Nocera, D. M. Tiede, H. Dau, W. Hillier, L. Wang, R. Amal, Artificial Photosynthesis as a Frontier Technology for Energy Sustainability, *Energy and Environmental Science*, 2013, vol. 6, p. 1074.

change technologies (such as AP) are globally deployed. These are found in Articles 14, 15 and 21.

Article 14 of the UDBHR relevantly provides

2.   Taking into account that the enjoyment of the highest attainable standard of health is one of the fundamental rights of every human being without distinction of race, religion, political belief and economic, or social condition, progress in science and technology should advance:

   (a)   access to quality health care and essential medicines, especially for the health of women and children, because health is essential to life itself and must be considered to be a social and human good;
   (b)   access to adequate nutrition and water;
   (c)   improvement of living conditions and the environment;
   (d)   elimination of the marginalisation and the exclusion of persons on the basis of any grounds;
   (e)   reduction of poverty and illiteracy.

Article 14 of the UDBHR goes beyond Article 66.2 of TRIPS (*Agreement on Trade-Related Aspects of Intellectual Property Rights*) by linking technology transfer explicitly to a list of five global public goods. In this sense Article 14 of the UDBHR may represent a potentially revolutionary step in the task of evolving international technology transfer norms. The provision links to international human rights law through the language of "fundamental rights".[68] More particularly, it expressly links progress in science and technology (such as AP) to the conceptual penumbra of the international right to health as well as economic, social and cultural rights. At the same time, the reference in 14.2 (a) to health being "a social and human good" emerges more from the language of bioethical discourse.[69]

   Article 14's reference to "progress" in science and technology supports the view that the focus in the UDBHR is on not just transfer of fully mature technology, but of knowledge transfer at the earliest stages of the research process. The reference in Article 14.2 (a) to progress in science and technology advancing access to essential medicines necessarily is likely to promote debate about the difference between "innovative" and "essential" medicines and whether each term should be defined by, for example, the operation of competitive markets or by scientific evidence of objectively demonstrated therapeutic benefit.[70]

Article 15 – sharing of benefits

1.   Benefits resulting from any scientific research and its applications should be shared with society as a whole and within the international community, in particular with

68   S. K. Sell, *Private Power, Public Law: The Globalization of Intellectual Property Rights*, Cambridge University Press, Cambridge 2003.
69   E. D. Pellegrino, Toward a Virtue Based Normative Ethics for the Health Professions, *Kennedy Institute of Ethics Journal*, 1995, vol. 5, pp. 253–277.
70   T. A. Faunce, The UNESCO Bioethics Declaration 'Social Responsibility' Principle and Cost-Effectiveness Price Evaluations for Essential Medicines, *Monash Bioethics Review*, 2005a, vol. 24, no. 3, pp. 10–19.

developing countries. In giving effect to this principle, benefits may take any of the following forms:

(a) special and sustainable assistance to, and acknowledgement of, the persons and groups that have taken part in the research;
(b) access to quality health care;
(c) provision of new diagnostic and therapeutic modalities or products stemming from research;
(d) support for health services;
(e) access to scientific and technological knowledge;
(f) capacity-building facilities for research purposes;
(g) other forms of benefit consistent with the principles set out in this Declaration.

2. Benefits should not constitute improper inducements to participate in research.

This provision appears considerably broader than the articulations of technology transfer norms in public international law. First of all, the concept of sharing benefits here expressed may permit, for example, evidence-based expert assessment of whether the technology being transferred is actually of cost-effective benefit to the developing nation, that is, a science-based evaluation of whether it represents true 'health innovation'. Second, the sharing of diagnostic and therapeutic modalities or products stemming from research (core aspects of technology transfer norms under public international law) is linked here with provision of and access to health services and care.[71] Third, technology transfer is here associated with principles requiring access to scientific and technological knowledge and capacity-building facilities for research purposes.

Article 21 – transnational practices – of the UDBHR provides

1. States, public and private institutions, and professionals associated with transnational activities should endeavour to ensure that any activity within the scope of this Declaration, undertaken, funded or otherwise pursued in whole or in part in different states, is consistent with the principles set out in this Declaration.
2. When research is undertaken or otherwise pursued in one or more states (the host State(s)) and funded by a source in another state, such research should be the object of an appropriate level of ethical review in the host state(s) and the state in which the funder is located. This review should be based on ethical and legal standards that are consistent with the principles set out in this Declaration.
3. Transnational health research should be responsive to the needs of host countries, and the importance of research contributing to the alleviation of urgent global health problems should be recognised.
4. When negotiating a research agreement, terms for collaboration and agreement on the benefits of research should be established with equal participation by those party to the negotiation.

71 J. H. Barton, New Trends in Technology Transfer: Implications for National and International Policy, *ICTSD Programme on IPRs and Sustainable Development*, 2007, ICTSD Issue Paper No 18. ICTSD.

5. States should take appropriate measures, both at the national and international levels, to combat bioterrorism and illicit traffic in organs, tissues, samples, genetic resources and genetic-related materials.

This provision takes up the unique challenge to the global normative architecture surrounding the activities of multinational corporate actors provided by Article 1 of the UDBHR. Article 1 provides:

2. This Declaration is addressed to states. As appropriate and relevant, it also provides guidance to decisions or practices of individuals, groups, communities, institutions and corporations, public and private.

Article 21.1 of the UDBHR, in this context, may be viewed as creating non-binding, best endeavours encouragement for multinational corporations to ensure that their health technology research, for example, conforms to principles such as those set out in Articles 14 and 15 of the UDBHR. Articles 21.2 and 21.3 taken together appear to define 'transnational research' as research undertaken or otherwise pursued in one or more state(s), but funded by a source in another state. This careful wording makes clear that the funding need not be provided by the state itself, but may derive, for instance, from private corporate sources, national or multinational in origin. Enunciation of the precautionary principle in relation to new developments in biotechnology and of the duty of humans to protect the environment for its own sake would be significant in this context. Yet arguably the second half of Article 21.3 goes beyond this, towards the evolution of a broader norm relating to transnational benefit in global health technology research in particular.

The principle stated in the second half of UDBHR Article 21.3, that states and public and private corporate actors should recognise the 'importance of research contributing to the alleviation of urgent global health problems' has similarly important policy implications for global health. It has implications, for example, for university research policies, which could encourage a shift in academic decision making about research directions at the commencement of public-funded basic science projects. It likewise has implications for licensing conditions and governance oversight of multinational corporations and the extent to which nations begin to move towards a science-based rather than market assessment (distorted by advertising and anti-competitive practices) approach to regulating technology innovations like to assist global health (such as AP).[72]

Intersections of UDBHR technology transfer with public international law and trade agreements

The UDBHR principles relevant to new technologies should be read in the context of how developing countries over the last few decades have implemented a variety of domestic policies to facilitate technology transfer from developed nations and multinational corporations and to encourage transnational benefit from new health technologies. These range from policies promoting science education, to funding for the creation and acquisition of

---

72 T. A. Faunce, Toward a Treaty on Safety and Cost-Effectiveness of Pharmaceuticals and Medical Devices: Enhancing an Endangered Global Public Good, *Globalization and Health*, 2006, vol. 2, pp. 5–15.

innovative technology, tax incentives for purchase of capital equipment and increased and enforced intellectual monopoly privileges (IMPs) (generally termed 'intellectual property rights' (IPRs)).[73] In the late 1970s, many developing countries sought in vain a Code of Conduct to regulate technology transfer under United Nations (UN) auspices.[74]

The *International Covenant on Civil and Political Rights* (ICCPR) and *International Covenant on Economic, Social and Cultural Rights* (ICESCR) (in Articles 2 and 2 (1) and 3 respectively) require states to take steps "individually and through international assistance and cooperation, especially economic and technical" to fulfil their human rights obligations in a manner that is non-discriminatory and responsive to the needs of the most vulnerable and marginalised groups. Also relevant could be the international human right to seek, receive and impart information which is part of the right to freedom of expression and the right to the enjoyment of the benefits of scientific progress (Article 19 ICCPR and Article 15(1) (b) ICESCR). Progressive realisation of the international human right to health, in this context, remains an important normative component of the *Universal Declaration of Human Rights* (UDHR) and ICESCR (Article 25 UDHR, Article 12 ICESCR). In the same category is the human right to share in scientific advancement and its benefits (Article 27 UDHR).[75] Presently nearly a hundred multilateral agreements refer to technology transfer, mostly as a 'transfer in' process by which developing countries seek to gain access to technical goods and know-how imported from the developed world.[76]

In the late 1990s, technology transfer was strategically incorporated into agendas in the World Trade Organization (WTO) that regulated technology as 'tradable commodity'.[77] The *Agreement on Trade-Related Aspects of Intellectual Property Rights* (known as TRIPS) in Article 7 noted that IPRs (IMPs) should contribute to the promotion of technological innovation and the transfer and dissemination of technology. Article 8.2 permitted countries to adopt "appropriate measures" to prevent the abuse of IPRs (IMPs) or "resort to practices" that "adversely affect the international transfer of technology". Furthermore, Article 66.2 of TRIPS addresses the issue of development, providing that:

> Developed country Members shall provide incentives to enterprises and institutions in their territories for the purpose of promoting and encouraging technology transfer to least-developed country Members in order to enable them to create a sound and viable technological base.

In 2001, WTO members established a Working Group on Trade and Technology Transfer to examine the relationship between trade and the transfer of technology to developing countries. In the same year, the TRIPS Council required developed-country members to submit detailed reports on the functioning in practice of the incentives provided to their enterprises for the transfer of technology in pursuance of their commitments under

73 G. Martin, C. Sorenson, T. Faunce, Balancing Intellectual Monopoly Privileges and the Need for Essential Medicines, *Globalization and Health*, 2007, vol. 3, pp. 4–8.

74 C. M. Correa, Can the TRIPS Agreement Foster Technology Transfer to Developing Countries?, in K. E. Maskus, J. H. Reichman (eds.), *International Public Goods and Transfer of Technology Under a Globalised Intellectual Property Regime*, Cambridge University Press, Cambridge 2005, pp. 227–256.

75 B. Toebes, Towards an Improved Understanding of the International Human Right to Health, *Human Rights Quarterly*, 1999, vol. 21, pp. 661–679.

76 K. E. Maskus, J. H. Reichman (eds.), *International Public Goods and Transfer of Technology Under a Globalized Intellectual Property Regime*, Cambridge University Press, Cambridge 2005.

77 J. H. Barton, *op. cit.*

Article 66.2. This reflected a long history of efforts by developing countries to enhance the relevance of the WTO for development, including the earlier, stalled proposal for a *Code of Conduct on Technology Transfer.*[78]

Licensing has become one of the major legal methods of technology transfer. It involves a permission granted by the patent owner to another to use the patented invention on agreed terms and conditions, while the patent owner continues to retain ownership of the patent. Licensing not only creates an income source for the patentee, but also establishes the legal framework for the transfer of the technology to developing nation researchers and engineers. A nation's power to regulate licensing practices that are abusive of technology transfer is contained in Article 40 of TRIPS. Technology transfer as a process often commences with innovations in academic institutions created with public funds. Successful technology transfer generally requires adaptive investments by local firms in technologies made available and affordable.[79]

Yet, a major stumbling block to the populations of developing countries gaining benefit from such initiatives remains that patent holders (often multinational corporations) view norms of technology transfer and transnational benefit as disproportionately cutting into their profits while adding to their costs.[80] This has resulted in peak non-governmental organisations and developing-nation stakeholders suggesting that norms of technology transfer under public international law have been deliberately shaped as 'soft law' best endeavours principles lacking the type of enforcement mechanism that IPRs (IMPs) gained under TRIPS (for instance trade sanctions upon breach of obligations). In practice, such stakeholders often claim, technology transfer norms under public international law tend to merely facilitate multinational corporations locating their production facilities within developing countries, to take advantage of the cheap labour or low-cost natural resources.[81]

This controversy provided part of the background to the adoption of the *Doha Declaration on TRIPS and Public Health* in 2003.[82] Paragraph 7 of the Declaration provides:

> We reaffirm the commitment of developed-country members to provide incentives to their enterprises and institutions to promote and encourage technology transfer to least-developed country members pursuant to Article 66.2.

Yet, developing nations continue to raise concerns at the Council for TRIPS about the lack of effective action by developed countries to comply with Article 66.2 of the TRIPS Agreement.

---

78 H. Ullrich, Expansionist Intellectual Property Protection and Reductionist Competition Rules; a TRIPS Perspective, in K. E. Maskus, J. H. Reichman (eds.), *International Public Goods and Transfer of Technology Under a Globalised Intellectual Property Regime*, Cambridge University Press, Cambridge 2005, pp. 726–757.

79 K. E. Maskus, K. Saggi, T. Puttitanun, Patent Rights and International Technology Transfer through Direct Investment and Licensing, in K. E. Maskus, J. H. Reichman (eds.), *International Public Goods and Transfer of Technology Under a Globalized Intellectual Property Regime*, Cambridge University Press, Cambridge 2005, pp. 265–281.

80 A. F. Holmer, M. C. Reif, J. S. Schwarting, R. A. Bohrer, T. A. Hayes, N. Rightor, W. K. Summers, J. Driscoll, J. M. Orient, C. Jacobs, D. J. Mehta, M. Angell, The Pharmaceutical Industry – To Whom Is It Accountable?, *NEJM*, 2000, vol. 343, pp. 1415–1417.

81 Susan K. Sell, *Private Power, Public Law: The Globalization of Intellectual Property Rights*, Cambridge University Press, Cambridge 2003, p. 83.

82 J. T. Gathii, The Legal Status of the Doha Declaration on TRIPS and Public Health Under the Vienna Convention on the Law of Treaties, *Harvard Journal of Law and Technology*, 2002, vol. 15, no. 2, pp. 291–317.

The World Intellectual Property Organization (WIPO) Development Agenda in its efforts to promote technology transfer refers chiefly to public international law instruments such as the ICCPR, ICESCR and TRIPS and makes little obvious reference to bioethics.[83]

The above discussion has highlighted tensions between developing and developed nations in this area. It has nonetheless demonstrated that norms of technology transfer and transnational benefit do have a legitimate and explicit place in core texts of public international law.[84] Those texts can be viewed as providing their own legitimacy for such norms in a positivist manner within the traditional framework of international law set out, as mentioned, in Article 38 of the *Statute of the International Court of Justice*. This becomes an important part of the conceptual background for technology transfer and transnational benefit norms in the UDBHR.

A comparison can be made here to the *Norms on the Responsibilities of Transnational Corporations and Other Business Enterprises with Regard to Human Rights*, which was adopted by the UN Sub-Commission on the Promotion and Protection of Human Rights in August 2003, after years of effort and deliberation. The legal status of the latter instrument is similar to the UDBHR in the sense that it is likely to influence, but has not acquired any formal status under public international law.[85] Although the *Norms on the Responsibilities of Transnational Corporations and Other Business Enterprises with Regard to Human Rights* is arguably a restatement of international legal principles about corporate social obligations, it remains controversial whether these types of norms targeting individual human beings and corporations (artificial entities, recognised as persons for some legal purposes) can be adequately explained within the traditional framework of public international law.[86] Public international law has established itself as a valuable but by no means sufficient system in which obligations related to international public health and technology transfer may be developed and debated.[87,88]

It could also be argued that there is no need to locate a definitive normative foundation in public international law now for the UDBHR or its technology transfer and transnational benefit provisions. Given its existing inchoate formal status under public international law, the UDBHR, like the *Norms on the Responsibilities of Transnational Corporations and Other Business Enterprises with Regard to Human Rights* could be viewed as a transition phase towards health norms under international 'hard' law such as those in the regional *European Convention on Human Rights and Biomedicine*. In force since 1997 (having acquired the requisite number of ratifications), this latter regional convention has a firm normative status under international law. Its status under customary international law is also strong, the European Court of Human Rights having taken it into account in dealing with cases where the relevant countries had not even ratified or signed the document.[89] It covers comparable

83 K. E. Maskus, J. H. Reichman, *op. cit.*

84 A. L. Taylor, Globalisation and Biotechnology: UNESCO and an International Strategy to Advance Human Rights and Public Health, *American Journal of Law and Medicine*, 1999, vol. 25, pp. 479–541.

85 D. Weissbrodt, M. Kruger, Human Rights Responsibilities of Business as Non-State Actors, in P. Alston (ed.), *Non-State Actors and Human Rights*, Oxford University Press, Oxford 2005 pp. 315–350.

86 D. Kinley, R. Chambers, The UN Human Rights Norms for Corporations: The Private Implications of Public International Law, *Human Rights Law Review*, 2006, vol. 6, pp. 447–497.

87 D. P. Fidler, Return of the Fourth Horseman: Emerging Infectious Diseases and International Law, *Minnesota Law Review*, 1997, vol. 81, pp. 771–775.

88 C. M. Correa, *op. cit.*

89 H. Nys, Towards an International Treaty on Human Rights and Biomedicine? Some Reflections Inspired by UNESCO's Universal Declaration on Bioethics and Human Rights, *European Journal of Health Law*, 2005, vol. 13, pp. 5–8.

matters relevant to technology transfer, such as equitable access to health care (Article 3) and scientific research (Chapter V). Thus principles of technology transfer and transnational benefit in the UDBHR, like those in the UDHR, may not only eventually be accepted as a part of customary international law, but as part of a similarly binding international convention or conventions.[90] This *International Convention on Bioethics and Human Rights* may have its own monitoring committee receiving states' reports, issuing general comments and receiving communications from individuals concerning breaches of such obligations.

'UNESCO Declaration on Photosynthesis and Human Rights'

We have seen the UNESCO *Universal Declaration on the Human Genome and Human Rights* and the UDBHR encourage the evolution of legal norms coherent with bioethics and human rights that challenge norms privileging corporate power. The next stage towards establishing the governance conditions for the Sustainocene could be a UNESCO *Universal Declaration on Photosynthesis and Human Rights*. Such a text could be the first to declare that photosynthesis, arguably a more important invention of biology than the human genome, as common heritage of not just humanity but all forms of life.

This would begin the normative process of protecting photosynthesis from excessive patents promoting inequitable or unsustainable use within the class of United Nations treaties involved with protecting the common heritage of humanity (such provisions cover, for instance, outer space,[91] the moon,[92] deep sea bed,[93] Antarctica[94] and world natural heritage sites[95]). Five core components are generally regarded as encompassing the common heritage of humanity concept under public international law. First, there can be no private or public appropriation; no one legally owns common heritage spaces or materials. Second, representatives from all nations must manage such resources on behalf of all (this often necessitating a special agency to coordinate shared management). Third, all nations must actively share with each other the benefits acquired from exploitation of the resources from the commons heritage region, this requiring restraint on the profit-making activities of private corporate entities and linking the concept to that of global public good. Fourth, there can be no weaponry developed using common heritage materials. Fifth, the commons should be preserved for the benefit of future generations.[96,97]

The claim for artificial photosynthesis and its core components to common heritage status would likely be at an inchoate stage initially. Probably the closest analogies involve

90 Council of Europe, *Convention for the Protection of Human Rights and the Dignity of Human Beings with Regard to the Application of Biology and Medicine.* ETS 164, Strasbourg.
91 United Nations, *Treaty on Principles Governing the Activities of States in the Exploration and Use of Outer Space, Including the Moon and Other Celestial Bodies*, art 1. Jan. 27, 1967, 18 U.S.T. 2410, 610 U.N.T.S. 205.
92 United Nations, *Agreement Governing Activities of States on the Moon and Other Celestial Bodies*, art. 1, Dec. 17, 1979, 18 I.L.M. 1434.
93 United Nations, *Convention on the Law of the Sea*, art. 1, para. 1, Dec. 10, 1982, 1833 U.N.T.S. 397.
94 Antarctic Treaty art. VI., Dec. 1, 1959, 12 U.S.T. 794, 402 U.N.T.S. 72.
95 UNESCO. World Heritage Convention. Available at: http://whc.unesco.org/en/conventiontext/.
96 J. Frakes, Notes and Comments: The Common Heritage of Mankind Principle and the Deep Seabed, Outer Space, and Antarctica: Will Developed and Developing Nations Reach a Compromise?, *Wisconsin International Law Journal*, 2003, vol. 21, p. 409.
97 A. Pardo, Whose Is the Bed of the Sea?, *Proceedings of the American Society of International Law*, 1968, vol. 62, p. 216.

claims that genetic diversity of agricultural crops,[98] plant genetic resources in general,[99] biodiversity[100] or the atmosphere[101] should be treated as not just areas of common concern but subject to common heritage requirements under international law.

The UNESCO *Universal Declaration on the Human Genome and Human Rights*, for example, in Article 1 declared that: "The human genome underlies the fundamental unity of all members of the human family, as well as the recognition of their inherent dignity and diversity. In a symbolic sense, it is the heritage of humanity." Article 4 states: "The human genome in its natural state shall not give rise to financial gains".[102] Other international law concepts coherent with such a principle include those that may declare the basic processes of photosynthesis a global public good,[103] an aspect of technology sharing obligations,[104] or those arising under the international right to health (set out for example in Article 12 of the United Nations *International Covenant on Civil and Political Rights*).[105] The UNESCO *Declaration on the Responsibilities of the Present Generations Towards Future Generations* expresses a concept of planetary common heritage that could encompass photosynthesis in Article 4:

> The present generations have the responsibility to bequeath to future generations an Earth which will not one day be irreversibly damaged by human activity. Each genera-tion inheriting the Earth temporarily should take care to use natural resources reason-ably and ensure that life is not prejudiced by harmful modifications of the ecosystems and that scientific and technological progress in all fields does not harm life on Earth.[106]

Planetary medicine is now a growing field in which the expertise of medical professionals is directed towards issues of global health and environmental protection, particularly includ-ing climate change.[107] A UNESCO *Declaration on Photosynthesis and Human Rights* could well support a global scientific endeavour on that topic and be promoted through domestic

---

98 C. Fowler. Biological Diversity in a North-South Context, in H. O. Bergersen, G. Parmann (eds.), *Green Globe Yearbook*, Oxford University Press, Oxford 1993, p. 33.

99 United Nations Food and Agricultural Organization (FAO), *International Undertaking on Plant Genetic Resources*. Art 1. Res 8/83, 1983.

100 M. Bowman, C. Redgwell, *International Law and Conservation of Biological Diversity: Kluwer Law International*, The Hague 1996. 33 at pp. 39–40.

101 United Nations, *Legal Status of the Atmosphere*, Para 1. UN Res 43/53 6 Dec 1988.

102 UNESCO, Universal Declaration on the Human Genome and Human Rights. Available at: http://www.unesco.org/new/en/social-and-human-sciences/themes/bioethics/human-genome-and-human-rights/.

103 I. Kaul, E. U. von Weisacker, O. R. Young, M. Finger (eds.), *Limits to Privatisation*, Earthscan, Lon-don, 2006, p. 311.

104 T. A. Faunce, H. Nasu, Three Proposals for Rewarding Novel Health Technologies Benefiting Peo-ple Living in Poverty: A Comparative Analysis of Prize Funds, Health Impact Funds and a Cost-Effectiveness/Competitive Tender Treaty, *Public Health Ethics*, 2008, vol. 1, no. 2, p. 146.

105 United Nations, *International Covenant on Economic, Social and Cultural Rights*. Adopted and opened for signature, ratification and accession by General Assembly resolution 2200A (XXI) of 16 December 1966, entry into force 3 January 1976. Available at: http://www.ohchr.org/EN/ProfessionalInterest/Pages/CESCR.aspx.

106 UNESCO, *Declaration on the Responsibilities of the Present Generations Towards Future Generations*, November 12, 1997. Available at: http://portal.unesco.org/en/ev.php-URL_ID=13178&URL_DO=DO_TOPIC&URL_SECTION=201.html.

107 T. McMichael, op. cit.

and international media as a defining symbolic endeavour of planetary nanomedicine.[108,109] One benefit of this for artificial photosynthesis researchers is that funding agencies respond indirectly to public and governmental national interest concerns and nanotechnology, despite its great promise, still has a problematic place in the popular imagination owing to safety issues. A Global Artificial Photosynthesis Project therefore represents an excellent opportunity to create high-profile awareness of nanotechnology as a positive contributor to overcoming major contemporary public health and environmental problems.

Conclusion: normative foundations in an emergent cosmopolitanism?

It has been argued that the UDBHR provisions on just and equitable technology transfer addressed to corporations are likely to play an important role in the world's governance transition from Corporatocene to Sustainocene. The UDBHR's role in this respect initially may be to influence public opinion and the moral views of policy makers, judges and legislators. The UDBHR's drafting history suggests that that text cannot readily be conceived as a codification on the public international law plane of extant bioethical principles. Similarly, only with some stretching of significance from the initial UDBHR preamble and unjustifiable polarisation of bioethics scholarship can it be considered a summary of principles emerging from a purely virtue-based bioethics normative tradition distinct from public international law.

One way of locating a solid normative foundation for technology transfer and transnational benefit norms in the UDBHR could be to focus on their link to well-established human rights and bioethical documents such as those mentioned in the preamble. Instruments containing technology transfer principles mentioned in the UDBHR preamble, such as the UDHR, the ICCPR and the ICESCR, are unambiguously part of the corpus of international public law through various components of the rule of recognition in Article 38 of the *Statute of the International Court of Justice*. The UDBHR itself, however, and bioethical texts such as the *Helsinki Declaration* are not generally accepted as part of customary international law under this positivist normative mechanism.

Another approach might seek to normatively found the UDBHR and its technology transfer and transnational benefit principles on the type of cosmopolitan normative thinking adopted by activists with a strong sense of conscience (for example those in non-governmental organisations (NGOs) such as *Medecins Sans Frontiers, Medact* or *Oxfam*). Such stakeholders in international civil society begin increasingly to strive for constant application of principles not derived from often dubious institutionalised methods of rule-making by which states acquire and maintain power, but from an emerging normative cosmopolitanism. Such a cosmopolitan normative foundation breaks with the seeming inevitability of nation states as instruments of global governance, replacing them with forms of liquid democracy in which all mature citizens can register to vote electronically (or pass their vote to someone individually considered more qualified) on legislation enacted in a parliamentary institution representing humanity as a whole.[110]

---

108 T. A. Faunce. *Nanotechnology for Sustainable Energy Conference*, sponsored by the European Science Foundation, Obergurgl, Austria, July 2010.
109 T. A. Faunce. *15th International Congress of Photosynthesis*, Beijing, August 2010.
110 T. A. Faunce, Developing and Teaching the Virtue-Ethics Foundations of Healthcare Whistle Blowing, *Monash Bioethics Review*, 2004, vol. 23, no. 4, pp. 41–55.

By conceding that a morally arbitrary boundary such as the boundary of the nation has a deep and formative role in our deliberations, we seem to be depriving ourselves of any principled way of arguing to citizens that they should in fact join hands across these other barriers.[111]

Such normative cosmopolitanism must nonetheless confront transitional difficulties associated with the contrary economic and political power of global corporate entities, the absence of formal citizenship status in a world government, and claims that cosmopolitan normative foundations promote an abstract, utopian sense of humanity that obscures family, cultural and community bonds and dangerously contradicts mass media influences and obligations to obey domestic laws (often despite how undemocratically they are manipulated by ruling elites).[112] There are influential stakeholders set to lose fortunes if large numbers of people decide to build upon global distribution of AP technology that gives them ready equitable access to fuel, food and fertilizer to diminish national and corporate allegiances in favour of local, globally linked associations and supporting universal ideals and social virtues such as justice, equity and respecting human dignity, but the equally significant non-anthropocentric virtue of environmental sustainability. The UDBHR will have a critical role to play in this normative transition to Sustainocene values.

References

Barton, John H. 2007, *New Trends in Technology Transfer: Implications for National and International Policy*, ICTSD Programme on IPRs and Sustainable Development, ICTSD Issue Paper No 18. ICTSD, Geneva.

Beer, Christian, Reichstein, Markus, Tomelleri, Enrico, Ciais, Philippe, 2010, Terrestrial Gross Carbon Dioxide Uptake: Global Distribution and Covariation with Climate, *Science*, 329: 834–838.

Blankenship, Robert E. 2002, *Molecular Mechanisms of Photosynthesis*, Oxford/Malden: Blackwell Science.

Bowman, Michael, Redgwell, Catherine 1996, *International Law and Conservation of Biological Diversity*, The Hague: Kluwer Law International.

Correa, Carlos M. 2005, Can the TRIPS Agreement Foster Technology Transfer to Developing Countries?, in *International Public Goods and Transfer of Technology Under a Globalized Intellectual Property Regime* (pp. 227–256), Maskus, K.E., Parmann, G. eds. Cambridge: Cambridge University Press.

Crutzen, Paul J. 2002, Geology of Mankind, *Nature*, 415: 23–26.

Crutzen, Paul J., Stoermer, Eugene 2000, The Anthropocene, *Global Change Newsletter*, 41: 17–18.

Esty, Daniel C. 2001, Bridging the Trade-Environment Divide, *Journal of Economic Perspectives*, 15: 113–130.

Faunce, Thomas Alured 2004, Developing and Teaching the Virtue-Ethics Foundations of Healthcare Whistle Blowing, *Monash Bioethics Review*, 23(4): 41–55.

Faunce, Thomas Alured 2005a, The UNESCO Bioethics Declaration "Social Responsibility" Principle and Cost-Effectiveness Price Evaluations for Essential Medicines, *Monash Bioethics Review*, 24(3): 10–19.

111 M. C. Nussbaum, Patriotism and Cosmopolitanism, *Boston Review*, October 1, 1994. Available at http://bostonreview.net/martha-nussbaum-patriotism-and-cosmopolitanism.

112 R. Fine, Taking the 'Ism' Out of Cosmopolitanism, *European Journal of Social Theory*, 2003, vol. 6, no. 4, pp. 451–470.

Faunce, Thomas Alured 2005b, Will International Human Rights Subsume Medical Ethics? Intersections in the UNESCO Universal Bioethics Declaration, *Journal of Medical Ethics*, 31: 173–178.

Faunce, Thomas Alured 2006, Toward a Treaty on Safety and Cost-Effectiveness of Pharmaceuticals and Medical Devices: Enhancing an Endangered Global Public Good, *Globalization and Health*, 2: 5–15.

Faunce, Thomas Alured 2007, *Who Owns Our Health: Medical Professionalism, Law and Leadership Beyond the Age of the Market State*. Sydney: University of New South Wales Press.

Faunce, Thomas Alured 2008, Toxicological and Public Good Considerations for the Regulation of Nanomaterial-Containing Medical Products, *Expert Opinion in Drug Safety*, 7(2): 103–106.

Faunce, Thomas Alured 2010a, *15th International Congress of Photosynthesis*, August 2010, Beijing.

Faunce, Thomas Alured 2010b, *Nanotechnology for Sustainable Energy Conference*, sponsored by the European Science Foundation, Obergurgl, Austria.

Faunce, Thomas Alured 2012, Future Perspectives on Solar Fuels, in *Molecular Solar Fuels Book Series: Energy* (pp. 506–528) Wydrzynski, Thomas John, Hiller, Warwick, eds., Cambridge UK: Royal Society of Chemistry.

Faunce, Thomas Alured 2012, *Nanotechnology for a Sustainable World: Global Artificial Photosynthesis as the Moral Culmination of Nanotechnology*, Cheltenham: Edward Elgar.

Faunce, Thomas Alured, Lubitz, Wolfgang, Rutherford, A. William, MacFarlane, Douglas, Moore, Gary F., Yang, Peidong, Nocera, Daniel G., Moore, Tom A., Gregory, Duncan H., Fukuzumi, Shunichi, Yoon, Kyung Byung, Armstrong, Fraser A., Wasielewski, Michael R., Styring, Stenbjorn 2013, Energy and Environment Policy Case for a Global Project on Artificial Photosynthesis, *Energy and Environmental Science*, 6(3): 695–698.

Faunce, Thomas Alured, Nasu, Hitoshi 2008, Three Proposals for Rewarding Novel Health Technologies Benefiting People Living in Poverty: A Comparative Analysis of Prize Funds, Health Impact Funds and a Cost-Effectiveness/Competitive Tender Treaty, *Public Health Ethics*, 1(2): 146–153.

Faunce, Thomas Alured 2013, Artificial Photosynthesis as a Frontier Technology for Energy Sustainability, *Energy and Environmental Science*, 6: 1074.

Fidler, David P. 1997, Return of the Fourth Horseman: Emerging Infectious Diseases and International Law, *Minnesota Law Review*, 81: 771–775.

Fine, Robert 2003, Taking the "Ism" Out of Cosmopolitanism, *European Journal of Social Theory*, 6(4): 451–470.

Fischer, Gunther, Schrattenholzer, Leo 2001, Global Bioenergy Potentials Through 2050, *Biomass Bioenergy*, 20: 151–155.

Fowler, Cary 1993, Biological Diversity in a North-South Context, in *Green Globe Yearbook* (pp. 33–41) Bergersen, H.O., Parmann, G., eds., Oxford: Oxford University Press.

Frakes, Jennifer 2003, Notes and Comments: The Common Heritage of Mankind Principle and the Deep Seabed, Outer Space, and Antarctica: Will Developed and Developing Nations Reach a Compromise?, *Wisconsin International Law Journal*, 21: 409.

Furnass, Bryan 2012, *From Anthropocene to Sustainocene: Challenges and Opportunities*. Public Lecture. Australian National University, Sidney, 21 March 2012.

Ganguly, Samrat 1999, The Investor-State Dispute Mechanism (ISDM) and a Sovereign's Power to Protect Public Health, *Columbia Journal of Transnational Law*, 38: 13–68.

Gathii, James T. 2002, The Legal Status of the Doha Declaration on TRIPS and Public Health Under the Vienna Convention on the Law of Treaties, *Harvard Journal of Law and Technology*, 15(2): 291–317.

Hammarström, Leif, Hammes-Schiffer, Sharon 2009, Artificial Photosynthesis and Solar Fuels, *Accounts of Chemical Research*, 42(12): 1859–1860.

Holmer, Alan F., Reif, Max C., Schwarting, Steven J., Bohrer, Robert A., Hayes, Thomas A., Rightor, Ned, Summers, William K., Driscoll, James, Orient, Jane M., Jacobs, Claude, Mehta,

Dilip J., Angell, Marcia 2000, The Pharmaceutical Industry-To Whom Is It Accountable?, *New England Journal of Medicine*, 343: 1415–1417.

Hoogwijk, Monique, Faaij, André, van den Broek, Richard, Berndes, Göran, Gielen, Dolf, Turkenburg, Wim 2003, Exploration of the Ranges of the Global Potential of Biomass for Energy, *Biomass Bioenergy*, 25: 119–123.

Kaul, Inge, von Weizacker, Ernest Ulrich, Young, Oran R., Finger, Matthias (eds.) 2006, *Limits to Privatization*, Earthscan Publications: London 311.

Kinley, David, Chambers, Rachel 2006, The UN Human Rights Norms for Corporations: The Private Implications of Public International Law, *Human Rights Law Review*, 6: 447–497.

Kumar, Ajay, Jones, David D., Hanna, Milford A. 2009, Thermochemical Biomass Gasification: A Review of the Current Status of the Technology, *Energies*, 2(3): 556–581.

Leslie, Mitch 2009, Origins: On the Origin of Photosynthesis, *Science*, 323: 1286–1287.

Lovelock, James E. 1991, *Gaia, the Practical Science of Planetary Medicine*, London: Gaia Books.

MacKay, David 2009, *Sustainable Energy-Without the Hot Air*, Cambridge: UIT.

Martin, Greg, Sorenson, Corinna, Faunce, Thomas Alured 2007, Balancing Intellectual Monopoly Privileges and the Need for Essential Medicines, *Globalization and Health*, 3: 4–8.

Maskus, Keith E., Saggi, Kamal, Puttitanun, Thitima 2005, Patent Rights and International Technology Transfer through Direct Investment and Licensing, in *International Public Goods and Transfer of Technology Under a Globalized Intellectual Property Regime* (pp. 265–281) Maskus, Keith E., Reichman, Jerome H., eds., Cambridge: Cambridge University Press.

McMichael, Tony 2002, The Biosphere, Health, and "Sustainability", *Science*, 297(5584): 1093.

Nocera, Daniel G. 2006, On the Future of Global Energy, *Daedalus*, 135, 112–115.

Nussbaum, Marta C., 1994, Patriotism and Cosmopolitanism, *Boston Review*, <http://boston review.net/martha-nussbaum-patriotism-and-cosmopolitanism>.

Nys, Herman 2005, Towards an International Treaty on Human Rights and Biomedicine? Some Reflections Inspired by UNESCO's Universal Declaration on Bioethics and Human Rights, *European Journal of Health Law*, 13: 5–8.

Pace, Ronald 2005, in *Artificial Photosynthesis: From Basic Biology to Industrial Application* (p. 13) Collings, A. and Critchley, C., eds., Weinheim: Wiley-VCH Verlag.

Pachauri, Rajendra, Reisinger, Andy (eds.) 2007, *Report of the Intergovernmental Panel on Climate Change*, IPCCC, Geneva, Switzerland.

Pardo, Arvid 1968, Whose Is the Bed of the Sea?, *Proceedings of the American Society of International Law*, 62: 216–229.

Parikka, Matti 2004, Global Biomass Fuel Resources, *Biomass Bioenergy*, 27: 613.

Pellegrino, Edmund D. 1995, Toward a Virtue Based Normative Ethics for the Health Professions, *Kennedy Institute of Ethics Journal*, 5: 253–277.

Pittock, Barrie A. 2009, *Climate Change. The Science, Impacts and Solutions*, 2nd edition, Collingwood: CSIRO Publishing.

Ragauskas, Arthur J., Williams, Charlotte K., Davison, Brian H., Britovsek, George, Cairney, John, Eckert, Charles A., Frederick Jr., William J., Hallett, Jason P., Leak, David J., Liotta, Charles L., Mielenz, Jonathan R., Murphy, Richard, Templer, Richard, Tschaplinski, Timothy 2006, The Path Forward for Biofuels and Biomaterials, *Science*, 484: 311.

Raustiala, Kal 2005, Form and Substance in International Agreements, *The American Journal of International Law*, 99(3): 581–614.

Rogner, Hals-Hogner 2004, United Nations Development World Energy Assessment. *Geneva: United Nations*, 5: 162.

Salamanca-Buentello, Fabio, Deepa L., Court, Erin B., Martin, Douglas K., Daar, Abdallah S., Singer, Peter A. 2005, Nanotechnology and the Developing World, *PloS Med.*, 2: e97.

Sartbaeva, Asel, 2008, Hydrogen Nexus in a Sustainable Energy Future, *Energy and Environmental Science*, 1: 79–85.

Sell, Susan K. 2003, *Private Power, Public Law: The Globalization of Intellectual Property Rights*, Cambridge: Cambridge University Press.

Simma, Bruno, Alston, Philip 1992, The Sources of Human Rights Law: Custom, Jus Cogens and General Principles, *Australian Yearbook of International Law*, 12: 82–102.

Steffen, Will, Crutzen, Paul J., McNeill, John R. 2007, The Anthropocene: Are Humans Now Overwhelming the Great Forces of Nature, *AMBIO: A Journal of the Human Environment*, 36(8): 614–621.

Steffen, Will, Persson, Asa, Deutsch, Lisa, Zalasiewicz, Jan, Williams, Mark, Richardson, Katherine, Crumley, Carole, Crutzen, Paul, Folke, Carl, Gordon, Line, Molina, Mario, Ramanathan, Veerabhadran, Rockström, Johan, Scheffer, Marten, Schellnhuber, Hans Joachim, Svedin, Uno 2011, The Anthropocene: From Global Change to Planetary Stewardship, *AMBIO: A Journal of the Human Environment*, 40(7): 739–761.

Stern, Nicholas 2007, *The Economics of Climate Change: The Stern Review: Cabinet Office HM – Treasury*, Cambridge, UK: Cambridge University Press.

Taylor, Allyn L. 1999, Globalisation and Biotechnology: UNESCO and an International Strategy to Advance Human Rights and Public Health, *American Journal of Law and Medicine*, 25: 479–541.

Toebes, Brigit 1999, Towards an Improved Understanding of the International Human Right to Health, *Human Rights Quarterly*, 21: 661–679.

Ullrich, Hanns 2005, Expansionist Intellectual Property Protection and Reductionist Competition Rules; a TRIPS Perspective, *Journal of International Economic Law*, 7(2): 401–430.

Weissbrodt, David, Kruger, Muria 2005, Human Rights Responsibilities of Business as Non-State Actors, in *Non-State Actors and Human Rights* (pp. 315–350) Philip Alston, ed., Oxford: Oxford University Press.

X   The report of the International Bioethics Committee on vulnerability

A review

Adèle Langlois

Introduction

The *Universal Declaration on Bioethics and Human Rights*, adopted by the United Nations Educational, Scientific and Cultural Organization (UNESCO) in 2005, contains 28 articles, 15 of which (3 to 17 inclusive) expound bioethical principles. To help states and other stakeholders to promote and uphold these principles, UNESCO's International Bioethics Committee produces reports explaining them in depth and advising on their implementation. To date, it has published on consent (2008), social responsibility and health (2010a), vulnerability (2011), traditional medicine (2013), non-discrimination and non-stigmatisation (2014) and benefit sharing (2015). In the Foreword to the first of these reports, on consent, Pierre Sané, then Assistant Director-General for Social and Human Sciences at UNESCO, emphasised that the principles in the Declaration are not 'abstract', but 'about real and pressing ethical issues that shape our daily lives'. Thus, 'Immediately after the adoption of the Declaration, IBC committed itself to contribute to the promotion of the Declaration by pursuing and deepening the reflection on the principles set forth therein' (UNESCO 2008: 5). This chapter analyses the International Bioethics Committee's report on vulnerability in the light of the broader bioethics literature on this subject. The report elaborates Article 8 of the *Universal Declaration on Bioethics and Human Rights*, 'Respect for human vulnerability and personal integrity':

> In applying and advancing scientific knowledge, medical practice and associated technologies, human vulnerability should be taken into account. Individuals and groups of special vulnerability should be protected and the personal integrity of such individuals respected.
>
> (UNESCO 2005)

The chapter draws not only on the report itself, but the author's observations of the drafting process and subsequent discussions at meetings of UNESCO's International Bioethics Committee and Intergovernmental Bioethics Committee in 2010 and 2011. The report makes a significant contribution to bioethical reflections on the concept of vulnerability by (a) broadening its application beyond the research context, to healthcare and biotechnology and (b) considering societal as well as individual means of addressing vulnerability. Yet for these means to be implemented by states, ethics committees and communities, more detailed examples and guidance will be required.

The concept of vulnerability in bioethics

In terms of bioethics codes and guidelines, the concept of vulnerability was first used explicitly in the 1979 Belmont Report (Levine et al. 2004: 45; Lange, Rogers and Dodds 2013: 334; ten Have 2015: 395). Before that, vulnerability was implicit. In the Nuremberg Code of 1947 and the original version of the World Medical Association's Ethical Principles for Medical Research Involving Human Subjects (Declaration of Helsinki 1964), for example, informed consent is presented as the means by which to rectify the vulnerability of research participants that is inherent in the paternalistic physician–patient relationship. The Belmont Report lists 'racial minorities, the economically disadvantaged, the very sick, and the institutionalized' as especially vulnerable in three ways: they lack the capacity to consent to research, they are more liable to be coerced or exploited and they face a greater risk of harm than others. If these difficulties cannot be mitigated, the vulnerable group should be excluded from research (Lange, Rogers and Dodds 2013: 334; Aix Scientifics 2014). The 1993 and 2002 guidelines on biomedical research of the Council for International Organizations of Medical Sciences (CIOMS) encapsulate these strands as 'a substantial incapacity to protect one's own interests', citing inability to consent and lack of means, power or resources as the potential sources of this incapacity (CIOMS 1993; CIOMS 2002: 18, 64). UNESCO's *Universal Declaration on Bioethics and Human Rights* (2005) expands these ideas by applying the concept to healthcare as well as research (ten Have 2015: 395). Despite this proliferation of guidelines, there remains lack of clarity about how vulnerability should be defined and understood (Schonfeld 2013: 191; Cheah and Parker 2015: 152). Some scholars even reject vulnerability as a bioethical concept, seeing it as too broad or vague to be useful (Levine et al. 2004: 45; Haugen 2010: 210; Lange, Rogers and Dodds 2013: 333–334; Schrems 2014: 831; Tavaglione et al. 2015: 106; ten Have 2015: 395–397). Yet the empirical fact of vulnerability does not disappear because it is difficult to conceptualise and thus needs to be addressed in both research and healthcare settings.

The term "vulnerability" derives from the Latin word meaning "wound" (*vulnus*) and hence implies susceptibility to physical harm (Levine et al. 2004: 47; Carter 2009: 859; Toader, Damir and Toader 2013: 936; Turner and Dumas 2013: 666; Kelly 2015: 479). Ten Have (2015: 396) charts how the application of vulnerability has broadened from the 1970s onwards, from being primarily concerned with physical and mental well-being, to include political, social, economic, cultural and environmental vulnerabilities or insecurities. Most scholarship on the application of vulnerability to ethical decision making divides this range into two spheres: the common vulnerability of the human condition and the special vulnerability of some people caused by internal or external factors (see, for example, Haugen 2010: 203; Rogers, Mackenzie and Dodds 2012: 12; Benatar 2013: 42; Turner and Dumas 2013: 665; Schrems 2014: 831; ten Have 2015: 397; Straehle 2016: 38). In terms of special vulnerability, those who are disabled, for example, might be considered to be intrinsically more vulnerable than others (thus the source of the vulnerability is internal), although some argue that disability is a social construct (hence the cause is externally generated) (Neufeldt and Mathieson 1995: 174; Wendell 1996: 57; Thompson, Bacon and Auburn 2015: 1328). In between is the view that certain inherent characteristics (such as disability) *may* render a person especially vulnerable under certain social conditions. Here the situation is the focus, rather than the individual themselves (Aultman 2014: 16; Schrems 2014: 838; Tavaglione et al. 2015: 104; ten Have 2015: 397–398).

Autonomy, or lack of it, plays a key role in determining vulnerability (ten Have 2014: 88; Siriwardhana 2015: 189). The 2008 version of the Helsinki Declaration, for example (which was functional at the time the UNESCO report on vulnerability was drafted), states, 'Some research populations are particularly vulnerable and need special protection. These include those who cannot give or refuse consent for themselves and those who may be vulnerable to coercion or undue influence' (Para. 19) (WMA 2008; ten Have 2016: 48).[1] There is an inverse relationship here, such that the less autonomy a person has, the more vulnerable they are considered to be (Haugen 2010: 203; Toader, Damir and Toader 2013: 938; Kelly 2015: 478; Straehle 2016: 36). This can lead to two opposite reactions: a positive, solidaristic one of caring for the needs of others (ten Have 2015: 403) or an abuse of power (Kelly 2015: 479). The solidaristic response is elicited by general as well as special vulnerability. In order for human beings to survive the vagaries of life, they have had to work together to overcome the physical vulnerabilities inherent to the human condition (the need for shelter, for example). In this regard, vulnerability has positive as well as negative outcomes: 'it is the basis for exchange and reciprocity between human beings' (Levine et al. 2004: 47; Rogers, Mackenzie and Dodds 2012: 32; Turner and Dumas 2013: 666; ten Have 2014: 89 [quoted]). Related to this is the idea that autonomy, too, is dependent on social interaction, as it is only in relationship with others that autonomy can be learned and exercised (ten Have 2015: 401).

Types of vulnerability

In ethics, the concept of vulnerability has often been applied to particular populations, to the extent that the term "vulnerable populations" is more commonly used than "vulnerability" itself (Siriwardhana 2015: 189; ten Have 2015: 396). Later versions of the Declaration of Helsinki, in particular, refer to vulnerable populations and communities (2008) and vulnerable individuals and groups (2013), but do not contain the word "vulnerability" (WMA 2008; WMA 2013). The 2002 CIOMS guidelines specify that research should be carried out on 'the least vulnerable necessary to accomplish the purposes of the research' (CIOMS 2002: 18). Vulnerable 'classes of people' include children, the elderly, junior members within a hierarchical group (nursing students or soldiers, for example) and:

> [R]esidents of nursing homes, people receiving welfare benefits or social assistance and other poor people and the unemployed, patients in emergency rooms, some ethnic and racial minority groups, homeless persons, nomads, refugees or displaced persons, prisoners, patients with incurable disease, individuals who are politically powerless, and members of communities unfamiliar with modern medical concepts.
>
> (CIOMS 2002: 64–66)

This type of categorisation of vulnerability has not evolved without criticism (Aultman 2014: 16; ten Have 2015: 397–398). Levine et al. (2004: 44–46) contend that the list of vulnerable groups has grown so long that the concept has 'lost force', as nearly everyone qualifies for at least one group. But the concept can still be exclusionary. Luna

---

1 Note, however, that the wording of the 2013 version of the Declaration of Helsinki has been revised substantially. It now refers to 'vulnerable groups and individuals' in a similar way to Article 8 of the *Universal Declaration on Bioethics and Human Rights* (2005) and states that they should 'receive specifically considered protection', but does not give any indication of source or type of vulnerability (WMA 2013).

and Vanderpoel (2013: 325) refer to those listed as vulnerable groups in various ethical codes and guidelines as the "usual suspects" from 'lower resources settings'. They argue that the categorical approach to vulnerability is both too broad, as it 'labels entire subpopulations as vulnerable' and too narrow, as it is 'blind to the inclusion of individuals from society who are from a more privileged background'. As Levine et al. (2004: 47) put it, 'There is much that puts research participants at risk beyond their membership in a "vulnerable" group.' Labelling can have a stigmatising effect. Furthermore, those stereotyped as powerless are denied agency when decisions are made on their behalf in paternalistic fashion, thus compounding their vulnerability in terms of lack of autonomy. As not everyone within the relevant group will be vulnerable to the same degree, as they are unique individuals, the blunt instrument of categorisation leads to 'equal treatment of unequal's' [sic]. When this happens, key ethical principles of respect for autonomy, justice and beneficence are subordinated to non-maleficence (Macklin 2003: 473; Levine et al. 2004: 47; Haugen 2010: 209–210; Lange, Rogers and Dodds 2013: 337; Toader, Damir and Toader 2013: 939; Schrems 2014: 833–835 [quoted]; Siriwardhana 2015: 189).

Single categorisations also imply a person's vulnerability is fixed rather than dynamic and may cause other vulnerabilities (such as poverty) to be missed, leading to 'unrecognized exploitation and harm' (Aultman 2014: 16). They fail to deal with the complexity of vulnerability. Luna and Vanderpoel (2013: 326) write:

> One problem of the categorical model of vulnerability is that a person or a group of persons can suffer different kinds of vulnerabilities; a complexity that is overlooked if whole groups of persons are merely defined as vulnerable. A consequence of the categorical model is a simplistic answer to a complicated problem.

This simplistic answer can in fact exacerbate vulnerability. "Playing safe", for example, by excluding all those considered vulnerable from research, may increase their isolation and stigmatisation (which goes against broader human rights principles of participation and empowerment) and may also render them more vulnerable to poor health, as drugs cannot be tested in, or tailored to, their context. Thus attempts to do no harm have the opposite effect and lead to the least sensitive – and thus arguably least useful – research being conducted on those who least need it (Macklin 2003: 478; Park and Grayson 2008: 1106; Carter 2009: 862; Haugen 2010: 210; Lange, Rogers and Dodds 2013: 335; Schrems 2014: 835; Kalabuanga et al. 2015: 3).

In drafting revised guidelines in 2015, CIOMS took account of these criticisms, as follows:

> In the past, groups of vulnerable persons were excluded from participation in research because this was considered the most expedient way of protecting these groups (for example children, women of reproductive age, pregnant women). As a consequence of such exclusions, information about the diagnosis, prevention and treatment of diseases in such groups of persons is now limited. This has resulted in a serious injustice. If information about the management of diseases is considered a benefit that is distributed within a society, it is unjust to deprive groups of persons of that benefit. The need to redress these injustices by encouraging the participation of previously excluded groups in basic and applied biomedical research is widely recognized.
>
> (CIOMS 2015: 4–5)

The draft guidelines also seek to avoid the 'traditional approach to vulnerability' of labelling 'entire classes of individuals as vulnerable' (CIOMS 2015: 45). They include the same list of groups likely to be vulnerable, but warn, 'researchers and research ethics committees must avoid making judgments regarding the exclusion of such groups based on stereotypes. One proposed mechanism that can be used to avoid stereotyping is community consultation, where feasible, before and during the conduct of the research' (CIOMS 2015: 46).

Application of the concept

Given the debates around definition, reach and so on, ten Have (2015: 395) claims that the academic literature on vulnerability 'does not make clear how vulnerability should be understood, interpreted, and applied'. Nevertheless, several scholars in recent years have devised more nuanced approaches to understanding vulnerability that recognise its intersectionality. Luna and Vanderpoel suggest employing a layered rather than categorical account of vulnerability.[2] Layers relating to cognitive impairment and lack of autonomy, for example, may interact with social circumstances involving stigmatisation and human rights abuses. Once the layers are identified, multiple approaches can be developed to avoid, minimise or otherwise address them. Lange, Rogers and Dodds (2013: 336–337) build on Luna's work to present a typology of (often) overlapping sources of vulnerability: inherent, situational and pathogenic. The inherent sources are inescapable in that they are fundamental to the human condition, but are dynamic in that their extent will depend on age, disability and so on. The situational sources depend on context and may be short- or long-term. The pathogenic sources emanate from 'dysfunctional social or personal relationships', based on discrimination, neglect or abuse. (Thus pathogenic sources could include discriminatory or neglectful exclusion of vulnerable people from research.) The typology enables researchers to take into account the 'entire constellation' of vulnerabilities that potential participants face and thus avoid increasing people's vulnerability by enacting well-intentioned but inappropriate protections.

Vulnerability has the potential to be an 'action-guiding principle', in terms of designating specific responsibilities and duties (Straehle 2016: 37), but this potential is yet to be fulfilled, according to some scholars. Aultman (2014: 16) has called for further guidelines 'and other protections' that deal better with the complexity of vulnerability than current ones which assign people to particular groups (neither the 2005 *Universal Declaration on Bioethics and Human Rights* nor the 2013 UNESCO report on vulnerability are cited). Levine et al. (2004: 46) and Lange, Rogers and Dodds (2013: 334) similarly argue that current guidelines state that autonomy should be respected and exploitation avoided, through ethics committees giving vulnerable groups special consideration, for example, but give little guidance and direction as to how this can be achieved in terms of concrete duties and actions. Siriwardhana (2015: 189) is critical of the focus on definitions and categories of vulnerability within the bioethics literature, which has drawn attention away from 'an all-important point: identifying practical and realistic ways to minimise harm and exploitation of at-risk, susceptible research populations'. This has resulted in a lack of guidance on mechanisms to reduce or prevent harm and exploitation, which he suggests could be

---

2 Here they draw on Luna's 2009 paper "Elucidating the concept of vulnerability: layers not labels", in *International Journal of Feminist Approaches to Bioethics*, (2: 131–139), which is much cited in the literature on vulnerability.

rectified by post-research ethics evaluation (to complement existing pre-research assess-ments) that would compare the 'ideal versus the real scenario' and share lessons learned by, *inter alia*, compiling a database of case studies and vignettes (Siriwardhana 2015: 188, 191, 194–197).

Turner and Dumas (2013: 668) make the point that, through technological innovation, what may seem like intractable and inherent vulnerabilities at present (disability or terminal illness, for example) may become eradicable. This will depend on access to treatment, how-ever, which may be denied due to social vulnerabilities: 'The task of bioethics is to address the problems of scarcity in societies of abundance and to consider the consequences of med-ical technology that will increase social inequality.'[3] Toader, Damir and Toader (2013: 939) also warn that biomedical progress, particularly in biotechnology, may create or worsen vulnerabilities in terms of injustice or discrimination, rather than decrease them. Benatar (2013: 45) argues that there has been 'an over-emphasis on science, technology, and new knowledge to solve our problems,' which he attributes to the "hyper-individualism" of neoliberal globalisation. As an alternative, he suggests developing international law levelled at populations rather than individuals, to tackle the negative social determinants of health, such as poor education and living conditions, which render people vulnerable (Benatar 2013: 42–43). ten Have (2015: 403–405) also calls for a systemic approach, to change the conditions that lead to inequalities of health, for example, rather than individual empower-ment through autonomous decision making.

Informed consent, though an important principle, is not sufficient to address multiple and complex sources of vulnerability (Macklin 2003: 479; Levine et al. 2004: 46; Luna and Vanderpoel 2013: 326). Any attempt to tackle the social, economic and political vulner-abilities exacerbated by globalisation will simply reflect, not transcend, the neoliberal world order if conducted at the individual rather than societal level. ten Have (2014: 89) claims:

[A]s long as bioethics does not critically examine the production of vulnerability itself it does not address the root of the problem. Framing vulnerability as a deficit of auton-omy not only presents part of the whole story but it also implies a limited range of options and actions.

Agency is nevertheless important. ten Have (2014: 91) goes on to recommend including vulnerable groups in policy formation and practice. Similarly, Park and Grayson (2008: 1106) have suggested that ethics committees should view the relationship between research-ers and participants as a partnership, rather than one of potential exploitation, particularly if participants are members of a "vulnerable population". The focus should be the betterment of society, rather than the individual. Rogers, Mackenzie and Dodds (2012: 12–13, 32), meanwhile, advocate a 'fuller account of vulnerability' that not only entails protection from harm but promotion of well-being through societal interventions such as education and welfare, which will in turn enlarge people's capacity for agency.

With regard to particular categories of people who *may* be vulnerable, recent studies have presented ideas on how to include rather than exclude them from research and also avoid crudely lumping people together in groups. In a paediatric clinical trial in a poor

---

3 Whilst several of the papers on vulnerability cited in this chapter do not make explicit reference to the *Universal Declaration on Bioethics and Human Rights* (2005), Turner and Dumas's article was writ-ten expressly to promote the 'multidisciplinary and pluralistic dialogue' for which the Declaration calls (Turner and Dumas 2013: 663).

community in Democratic Republic of Congo, for instance, patients and their guardians who were unable to read required their informed consent to be confirmed by a literate witness. This not only curtailed their autonomy, but also left them vulnerable to the undue influence by the witness, whom they assumed better understood the research. Furthermore, as healthcare was not readily available in the community in which the research took place, there was competition to participate, irrespective of compatibility with the enrolment criteria (Kalabuanga et al. 2015: 4–5). Kalabuanga et al. (2015: 4–5) recommend in-depth community engagement and benefit sharing to address such issues. Illiteracy levels are particularly high among elderly people in Africa, which may compound other factors which put them at risk, such as 'health, functional status, chronic illness, and financial circumstances' (Lekalakala-Mokgele and Adejumo 2013: 501). In these circumstances using graphic symbols may be appropriate, whilst ensuring that the participant is not patronised. Lekalakala-Mokgele and Adejumo (2013: 500) further stipulate that 'a consent form should be viewed as a continuous process rather than an outcome to be achieved due to the changeable nature of the older persons' competence over time'. Low literacy levels can also be a barrier to participation in rich countries, among minorities, for example. Rogers and Lange (2013: 2143) advocate designing recruitment materials in relevant languages, as 'once effective communication is achieved, there can be no suggestion that minority groups lack capacity to consent'.

Pregnant women and children are two (wide) categories of people who have been persistently excluded from research. Barring research on pregnant women has not only prevented important research on pregnancy from taking place, but has also excluded women more generally from research, as those of child-bearing age have been ruled out of projects on the grounds that they might conceive during them (Schonfeld 2013: 190). Schonfeld (2013: 204) suggests an agency-centred approach to resolving this problem, through more research on the choices that women make with regard to research participation and relative risk. Recognising and encouraging agency are also at the heart of recent practical proposals for providing more opportunities for children to participate in research. Carter (2009: 858) advocates 'child-led' ethics committees, as these could provide an interesting and valuable contribution to adults' understanding of the ethical and governance issues that exercise children, thereby accommodating their 'strengths, expertise and capacities'. Ho, Reis and Saxena (2015: 245), drawing on discussions at the 10th Global Summit of National Bioethics Committees in 2014, also argue that children should help develop research projects. At the individual level, Cheah and Parker (2015: 151, 157) refute the blanket approach that designates all children as lacking sufficient capacity to consent, on the grounds that 'vulnerability is context- and study-specific'. Mature children in poor communities, for example, are often denied the opportunity to take part in research, even when they wish to, because their parents are unavailable (Cheah and Parker 2015: 151). Cheah and Parker (2015: 162) recommend waiving the requirement for parental consent in such circumstances, provided that other safeguards are met, such as approval by an ethics committee and use of accessible information formats and that the child's decision is judged voluntary, independent, mature and competent.

The report of UNESCO's International Bioethics Committee on vulnerability

The Report of the International Bioethics Committee of UNESCO on the Principle of Respect for Human Vulnerability and Personal Integrity consists of two main parts: a review of the determinants of "special vulnerability" and a series of case studies of vulnerability in different settings (UNESCO 2013). It was drafted by a Working Group, with input from the International Bioethics Committee (IBC) and the Intergovernmental Bioethics Committee (IGBC) at their various meetings from 2008 to 2011.[4] The final version was presented at the IGBC meeting in September 2011.[5] It represents two aims of the IBC: 'paving the way for a broader reflexion' and 'indicating possible lines of action' (UNESCO 2013: 5). The case studies, in particular, are intended to be used as a basic educational tool, to be adapted to the context of a country or region (personal observations, IBC meeting, May 2011). They are illustrative rather than exhaustive: 'a useful template for further discussion and development' (UNESCO 2013: 37). The report builds on the Universal Declaration on Bioethics and Human Rights (2005), thus fulfilling the IBC's mandate in terms of interpreting the Declaration in practicable ways, to aid implementation by states. For instance, the introduction avows, 'Article 8 of the Declaration entails both a "negative" duty to refrain from doing something and a "positive" duty to promote solidarity and to share the benefits of scientific progress' (UNESCO 2013: 9); this duty to share benefits is not immediately obvious from the wording of the article itself. The conclusion of the report draws on Article 1 of the Universal Declaration on Bioethics and Human Rights (2005)[6] to make clear that the responsibility to address special vulnerability lies not just with states, but also 'individuals, groups, communities, institutions and corporations, public or private' (UNESCO 2013: 37). Many of the case studies' "remedies" for the vulnerabilities they outline are targeted specifically at states, however (UNESCO 2013).

In presenting a preliminary draft of the report to the IBC at its October 2010 meeting, a member of the Working Group broke Article 8 of the *Universal Declaration on Bioethics and Human Rights* (2005) into four parts:

1   Human vulnerability should be taken into account.
2   In applying and advancing scientific knowledge, medical practice and associated technologies.
3   Individuals and groups of special vulnerability should be protected.
4   And the personal integrity of such individuals respected.

They explained that (1) is not the main target of Article 8, but that it provided a background to the Working Group's reflection. The argument put forward is very similar to ten Have's: vulnerability is a central feature of humankind, which means that humans have a duty of solidarity and assistance towards each other, as all humans have the potential to suffer. By contrast, God does not need our solidarity because He is not vulnerable or in need of help, though He asks for our respect and we have duties towards Him. The main target of Article 8, however, is the circumstances of (2). The Working Group's purpose was

4   Namely, preliminary reflections at the IBC meetings at UNESCO headquarters in Paris in 2008 and in Mexico in 2009, followed by more in-depth discussions of drafts at the IBC and joint IBC-IGBC meetings in Paris in 2010 and the IBC meeting in Azerbaijan in 2011. The author attended the 2010 and 2011 meetings, as well as the September 2011 IGBC meeting, as an observer.
5   Although the report was finalised in 2011, the downloadable version on the UNESCO website is dated 2013.
6   Article 1 states, 'This Declaration is addressed to States. As appropriate and relevant, it also provides guidance to decisions or practices of individuals, groups, communities, institutions and corporations, public and private' (UNESCO 2005).

to set out normative standards to foster an awareness of our duties to the people in (3), as we are not all equally vulnerable. These had to go beyond philosophical statements about the essence of human beings to be useful and meaningful for people of special responsibility and therefore respect their personal integrity, entailing a shift from a rights-based approach to bioethics to a more duties-based reflection (4) (personal observations, IBC meeting, October 2010).

The preliminary draft of the report contained several references to academic papers by prominent ethicists such as Ruth Macklin, Samia Hurst and Solomon Benatar (UNESCO 2010b). After feedback from the IBC, however, it was decided to remove these, to make the report a more practical than scholarly document and ensure it did not appear to represent one philosophical tradition or culture (personal observations, IBC meeting, October 2010). As Donald Evans, then Chair of the IBC, explained to the IGBC at its September 2011 meeting, the intended primary audience for the report comprises states, institutions and individual stakeholders, not academics. Thus it was important to try to make the report accessible, gripping and operationalisable (personal observations, IGBC meeting, September 2011). In earlier meetings both IBC and IGBC members had stressed that the document needed to have a practical rather than philosophical bent, with some requesting that it include actionable recommendations for ethics committees and governments (personal observations, IBC and joint IBC-IGBC meetings, October 2010 and IBC meeting, April/May 2011).

Relatedly, the report deliberately avoids defining vulnerability, for fear of doing so too narrowly or widely or waxing philosophical at the expense of utility. The report qualifies the omission by stating, 'In most cases, however, it is relatively easy to recognise vulnerability when it arises: something fundamental is indeed at stake' (personal observations, IGBC meeting, September 2011; UNESCO 2013: 13). This is potentially problematic, as those who are most vulnerable might not be easily seen, precisely because they are vulnerable and thus hidden from society. The decision to jettison a definition was not unanimously supported by IBC and IGBC members. Whilst some welcomed the avoidance of deep philosophical wrangling, several others felt that the exclusion weakens the report, as a reference definition would be a useful addition to the bioethics canon and, without one, the conceptualisation is too vague and thus potentially counter-productive, as it could engender misunderstandings. A definition might also have helped countries write guidelines on tackling vulnerability (personal observations, IBC and joint IBC-IGBC meetings, October 2010).

The report focuses on vulnerability in three contexts: healthcare, research and emerging biomedical technologies. This reflects Article 1 of the *Universal Declaration on Bioethics and Human Rights*, which specifies that the Declaration is designed to address 'ethical issues related to medicine, life sciences and associated technologies' (UNESCO 2005). As much of the literature on vulnerability is concerned with research settings, with relatively little attention given to medical treatment and healthcare (Tavaglione et al. 2015: 99), the inclusion of healthcare is welcome. In line with the Working Group's mandate as described above, the report focuses on special vulnerabilities. Article 8's recognition of vulnerability as 'an essential feature of human nature' notwithstanding (UNESCO 2013: 5, 13). The report lists the causes of these special vulnerabilities as 'personal disability, environmental burdens or social injustice' and draws a distinction between 'natural' and 'context-related' vulnerabilities (UNESCO 2013: 5). Special vulnerabilities are split into two 'fundamental' categories: (a) disabilities, disease and limitations (such as age) 'imposed by the stages of human life' and (b) social, political and environmental. The former are considered to be naturally occurring, whilst the latter are declared more complex as they involve 'the fundamental matter of justice'. The first category is determined according to personal characteristics in a way that has been criticised by Luna and others (see above); children, for example, are 'assumed' to be inherently vulnerable, so social context is irrelevant. The second category, however, is determined according to situational and pathogenic sources, such as poverty and discrimination (UNESCO 2013: 14–15). Where the report differs from the Lange et al. typology is that the possibility that these categories might intersect is not explicitly recognised; for instance, examples of grounds for marginalisation are listed, such as ethnicity, race and migrant status, but disability is not included (UNESCO 2013: 15).

The report also clearly makes the inverse link between vulnerability and autonomy that is so prevalent in the academic literature, stating that its focus is the special vulnerabilities and conditions that 'impinge upon the capacity to live as free and autonomous individuals' (UNESCO 2013: 5). Several IBC members emphasised the importance of this link during discussions of draft reports, one pointing out that the common factor in all the case studies presented at the October 2010 IBC meeting was that people could not say 'no' (personal observations, IBC meeting, October 2010). Like those scholars who have traced the meaning of "vulnerability" to its Latin roots, the report uses the word "wound", not only in the physical sense but also to describe the impact of denial of freedom on human identity (UNESCO 2013: 13). Nevertheless, the introductory remarks also recognise other causes and consequences of vulnerability, echoing Benatar's concerns about inequality as a source of injustice and vulnerability, particularly in the context of uneven access to medical and technological advances (UNESCO 2013: 5). This dual focus on autonomy and equality is maintained throughout the report (UNESCO 2013).

Like ten Have, the report recognises that vulnerability is a necessary trigger of societal progress, seeing efforts to address harmful circumstances as fundamental obligations of human beings and 'a prerequisite of human flourishing' (UNESCO 2013: 13). In terms of practical implementation, the report obfuscates whether states and other actors are to 'address' special vulnerabilities and their determinants (UNESCO 2013: 10), or simply take them into account, as per Article 8 of the *Universal Declaration on Bioethics and Human Rights* (2005). This relates to the broader debate about whether vulnerability can be eliminated, or only mitigated. Whilst the vulnerability inherent to the human condition cannot, by definition, be completely eliminated, special vulnerabilities that can be attributed to unequal social structures, for example, can be. As intimated by Benatar and ten Have, it is surely more effective in the long term to tackle the source of the vulnerability than to alleviate its effects.

In the set of case studies on vulnerability in the healthcare setting, there is an implicit recognition that the sources of vulnerability may be deeply structural. For instance, the remedy proposed to address the lack of antiretrovirals in developing countries (case study III.1.1), which can lead to a poor quality of life and untimely death, includes 'international solidarity' to encourage better healthcare provision on the part of states, which are to 'intervene directly by providing adequate health education and access to available therapies'. Similarly, with regard to access to essential tests and therapies more generally (III.1.2), the suggested remedy is 'The availability of appropriate healthcare resources to meet the needs of the patient population irrespective of ability to pay'. Yet there is no indication of how these remedies are to be effected (UNESCO 2013: 19), which will require a move away from market-based, neoliberal health policies (ten Have 2014: 90; ten Have 2015: 405). That the solution will involve a redistributive intervention on behalf of the state is made explicit in the remedy for the unfair allocation of healthcare resources (III.2.2): 'States should have in place a robustly resourced healthcare system that fairly and without discrimination provides adequate care to all citizens' (UNESCO 2013: 21). But how poorer states are to resource such healthcare systems is not addressed (unless the international solidarity of III.1.1 is assumed, perhaps through the formation of the alliances and networks ten Have (2014: 91) deems necessary to deal with 'global threats'). The remedy for poor health among migrants (II.1.3) is a little more specific and arguably, therefore, more useful: 'social integration of migrant individuals and communities in the mainstream, better and more targeted education about healthcare risks and ease of access to healthcare professionals' (although how these measures are to be resourced remains implicit) (UNESCO 2013: 20).

Lack of access to healthcare and resources also lie at the heart of several of the case studies on vulnerability in human participant research and in the development and application of emerging technologies, as institutions, doctors and patients are more likely to enroll in research if this appears to be the only means by which they can, respectively, access research funds (institutions), medicines for their patients (doctors) or healthcare for themselves or their family (patients) (see IV.1 on "double standard" research, IV.2 on equivocal donations, IV.4 on social responsibility, IV.5 on lack of research on tropical diseases and V.2 on unfair pressure; UNESCO 2013, 25–27, 31–32). In the example of women whose only means of accessing in vitro fertilization treatment (IVF) is to allow the clinic to harvest surplus eggs for use by other women (V.2), the report recommends 'stricter licensing, oversight, monitoring and evaluation of clinics offering these services' (UNESCO 2013: 32). This might help to prevent exploitation of women unable to afford treatment, but would do little to address inequitable access. Thus, in this instance, the report fails to deal with the challenge highlighted by Benatar and Turner and Dumas of developments in biotechnology leading to increased social inequality.

In the context of biomedical research, the report cites the CIOMS (2002) guidelines' stipulation that the involvement of vulnerable people in research needs to be specially justified, but also recognises that exclusion from research can increase vulnerability. The example of children being prescribed potentially inappropriate drugs that have only been tested on adults is given (UNESCO 2013: 16, 20). Whereas the academic literature and the draft revisions to the CIOMS guidelines recommend including potentially vulnerable individuals and groups in decision making about research to mitigate the potential for exploitation (see, for example, Carter (2009), ten Have (2014), Ho, Reis and Saxena (2015), Kalabuanga et al. (2015) and CIOMS (2015)), this is not a feature of any of the report's proposed remedies for vulnerability in human participant research. Suggested actions are directed at ethics committees, corporations, regulators and states, but not would-be research participants. Governments are to 'take responsibility *for* their citizens by developing policies that give priority *to* vulnerable communities with the aim of improving *their* quality of life' (UNESCO 2013: 25–27; emphasis added). The vulnerable communities appear to be assigned the role of passive partner in these endeavours. 'Proper consultation' with social groups is recommended in the section on emerging technologies, however, for genetic research that could lead to stigmatisation of said groups (case study V.1 on stigmatisation; UNESCO 2013: 31).

The report does engage with the issue of multiple vulnerabilities, the conclusion stating, 'It must be accepted that situations of vulnerability seldom exist in isolation' (UNESCO 2013: 37). This is done in a potentially problematic way with regard to gender, however. Women and girls are given 'special attention' in the report: 'Female cases are prominent as they are particularly exposed to the whole range of the social, cultural, economic, educational and political determinants of vulnerability' (UNESCO 2013: 5–6). In the case study on social vulnerability in human participant research (IV.4), 'being female' is listed as a contextual factor that might generate social vulnerability, without any reference to intersectionality or the type of layered approach to situation and context recommended by Luna and Vanderpoel.[7] Oddly, although marginalisation on the grounds of race and ethnicity is also listed as a possible factor, the suggested remedy is poverty alleviation (again, with no indication of how this is to be achieved) and 'strict limitations [not bans] on the use of potentially coercive incentives', rather than measures to stop discrimination (UNESCO 2013: 27). Relatedly, the problem of social hierarchies affecting autonomous decision making identified by Kalabuanga et al. (2015: 4) is not substantively addressed, except in relation to the doctor–patient relationship (UNESCO 2013: 15, 26).

Conclusion

The IBC report on vulnerability broadens the application of the concept of vulnerability beyond biomedical research, to healthcare settings and developments in biotechnology. This reflects UNESCO's approach to bioethics more generally, as espoused in its 2005 *Universal Declaration on Bioethics and Human Rights*. It also answers recent calls from the academic bioethics community for these areas to receive more attention. In terms of how to address vulnerability, the report also broadens the 'range of options and actions' (ten Have 2014: 90) beyond increasing individual autonomy, to incorporate societal level actions and interventions. Some of these remedial actions are sketched in such general terms, however, that they might be construed as of limited use to policy makers at best and, at worst, as wishful thinking. Such suggestions add little to existing human rights and human security discourses concerning the obligations of states for the economic and social welfare of their citizens and other residents. The recommendations for ethics committees, also, are unlikely to be detailed enough to meet the need for more concrete guidance identified by Levine et al., Aultman, Straehle and Lange, Rogers and Dodds, as well as IBC and IGBC members. Thus whether the report is sufficiently innovative or radical to fill some of the practical gaps identified in the bioethics literature is questionable, even if it recognises that the root causes of vulnerability in all three settings – healthcare, research and technological development – lie in social inequality, as highlighted by Benatar, ten Have and others.

Another indication that the report might not be having the desired impact is that it is cited in relatively few of the articles published on vulnerability after the release of its final version in 2011. This might equally be due to the need for further dissemination, however. Despite the best efforts of the Bioethics Programme's secretariat, not all of the bioethics community are more than peripherally aware of UNESCO's bioethics activities (Langlois 2013: 72–74, 103–104). During discussions of the preliminary draft report on vulnerability in 2010, IGBC delegates from Kenya and Venezuela emphasised that access to information, education and empowerment are vital to combatting vulnerability. Thus it is not just ethicists (whether philosophers, or ethics committee members, or both) who need to be reached, but the public at large (personal observations, joint IBC-IGBC meeting, October 2010). The

---

7 Luna and Vanderpoel (2013: 326) write, 'For example, being a woman does not *per se* entail vulnerability. However, a woman living in a country that does not recognize or is intolerant of *reproductive rights* would acquire a layer of vulnerability. In turn, an educated and resourceful woman in that same country can overcome some of the consequences of the denial of reproductive rights; however, a *poor woman* in that country acquires another layer of vulnerability. Moreover, an *illiterate* poor woman in that situation acquires yet another layer. And, if this woman is an *immigrant, undocumented* or belongs to an *aboriginal community*, she will acquire increasing layers of vulnerabilities and would suffer proportionately under these overlapping layers.'

report recognises this point in the conclusion: 'lower levels of education always predict higher levels of vulnerability' (UNESCO 2013: 37). Several IBC members had requested this at their 2011 meeting, especially given UNESCO's key role in education (personal observations, IBC meeting, April/May 2011). One option might be to enhance the case studies in the report and develop them into educational materials, to form a case study database of the type recommended by Siriwardhana (the Bioethics Programme has already produced casebooks on *Human Dignity and Human Rights* and *Benefit and Harm* (UNESCO 2016).

There are some other possible avenues for UNESCO to expand its work on vulnerability. Disaster relief and research are areas that require attention. 'Natural' disasters serve to exacerbate existing vulnerabilities, having vastly different impacts on people's health and welfare, depending on social context (Chung and Hunt 2012: 197–199), whilst the 'shifting moral landscape' that researchers of disasters face renders the vulnerabilities that they might exacerbate through their research particularly dynamic and complex (Eckenwiler et al. 2015: 654–655). Neither of these contexts features in the case studies in the IBC report on vulnerability, despite some IGBC members arguing for their inclusion (personal observations, joint IBC-IGBC meeting, October 2010). Another option might be to consider the vulnerability of non-human animals. Although humans are the focus of the *Universal Declaration on Bioethics and Human Rights* (and therefore of the IBC, IGBC and Bioethics Programme), Kelly (2015: 480) makes the point that neither the dependency of newborns, nor the capacity to wound or be wounded, are unique to humans. Thus the vulnerability that is inherent to the human condition is also inherent to all living creatures. This raises questions for what makes for ethical relations between humans and non-human animals. In terms of exploitation, for example, Kelly (2015: 481) notes, 'Some bodies are made vulnerable for the sake of the prosperity of others. In relation to human bodies and human institutions, nonhuman animal bodies are prime examples of vulnerable bodies, particularly those raised on factory farms for human consumption.' Finally, through its cultural arm, UNESCO could explore theological perspectives on vulnerability, which have so far been lacking (Haugen 2010: 211).

Acknowledgements

The research for this chapter was supported by the Faculty of Health, Life and Social Sciences (now the College of Social Science) Research Fund at the University of Lincoln and a Wellcome Trust Biomedical Ethics Programme Small Grant (ref: 096024).

References

Aix Scientifics 11 November 2014, *World Medical Association Deklaration von Helsinki 1964*, <www.aix-scientifics.co.uk/en/_helsinki64.html> accessed 02-08-2017.

Aultman, Julie 2014, Vulnerability: Its Meaning and Value in the Context of Contemporary Bioethics, *The American Journal of Bioethics*, 14(12): 15–17.

Benatar, Solomon R. 2013, Global Health, Vulnerable Populations, and Law, *Journal of Law, Medicine and Ethics*, 41(1): 42–47.

Carter, Bernie 2009, Tick Box for Child? The Ethical Positioning of Children as Vulnerable, Researchers as Barbarians and Reviewers as Overly Cautious, *International Journal of Nursing Studies*, 46(6): 858–864.

Cheah, Phaik Yeong, Parker, Michael 2015, Are Children Always Vulnerable Research Participants?, *Asian Bioethics Review*, 7(2): 151–163.

Chung, Ryoa, Hunt, Matthew R. 2012, Justice and Health Inequalities in Humanitarian Crises: Structured Health Vulnerabilities and Natural Disasters, in *Health Inequalities and Global Justice*, (pp. 197–212) Lenard, Patti Tamara, Straehle, Christine., eds., Edinburgh: Edinburgh University Press.

CIOMS)1993, *International Ethical Guidelines for Biomedical Research Involving Human Subjects*, <www.codex.uu.se/texts/international.html> accessed 14-03-2016.

CIOMS 2002, *International Ethical Guidelines for Biomedical Research Involving Human Subjects*, Geneva.

CIOMS 2016, *International Ethical Guidelines for Health-related Research Involving Humans*, <https://cioms.ch/> 02-08-2017.

Eckenwiler, Lisa, Pringle, John, Boulanger, Renaud, Hunt, Matthew 2015, Real-Time Responsiveness for Ethics Oversight during Disaster Research, *Bioethics*, 29(9): 653–661.

Haugen, Hans Morten 2010, Inclusive and Relevant Language: The Use of the Concepts of Autonomy, Dignity and Vulnerability in Different Contexts, *Medicine, Health Care, and Philosophy*, 13(3): 203–213.

Ho, Calvin W.L., Reis, Andreas, Saxena, Abha 2015, Vulnerability in International Policy Discussion on Research Involving Children, *Asian Bioethics Review*, 7(2): 230–249.

Kalabuanga, Marion, Ravinetto, Raffaella, Maketa, Viki, Lutumba, Pascal 2016, The Challenges of Research Informed Consent in Socio-Economically Vulnerable Populations: A Viewpoint from the Democratic Republic of Congo, *Developing World Bioethics*, 16: 64–69.

Kelly, Oliver 2015, Witnessing, Recognition, and Response, *Philosophy and Rhetoric*, 48(4): 473–493.

Lange, Marc M., Rogers, Wendy, Dodds, Susan 2013, Vulnerability in Research Ethics: A Way Forward, *Bioethics*, 27(6): 333–340.

Langlois, Adèle 2013, *Negotiating bioethics: the governance of UNESCO's bioethics programme*, London: Routledge.

Lekalakala-Mokgele, Eucebious, Adejumo, Oluyinka 2013, Conducting Research with African Elderly Persons: Is Their Vulnerability a Concern to Researchers?, *African Journal for Physical, Health Education, Recreation and Dance*, 19(2): 496–504.

Levine, Carol, Faden, Ruth, Grady, Christine, Hammeschmidt, Dale, Eckenwiler, Lisa, Sugarman, Jeremy 2004, The Limitations of "Vulnerability" as a Protection for Human Research Participants, *The American Journal of Bioethics*, 4(3): 44–49.

Luna, Florencia, Vanderpoel, Sheril 2013, Not the Usual Suspects: Addressing Layers of Vulnerability, *Bioethics*, 27(6): 325–332.

Macklin, Ruth 2003, Bioethics, Vulnerability, and Protection, *Bioethics*, 17(5–6): 472–486.

Neufeldt, Aldred H., Mathieson, Ruth 1995, Empirical Dimensions of Discrimination against Disabled People, *Health and Human Rights*, 1(2): 174–189.

Park, Stephanie S., Grayson, Mitchell H. 2008, Clinical Research: Protection of the "Vulnerable"?, *Journal of Allergy and Clinical Immunology*, 121(5): 1103–1107.

Rogers, Richard, Lange, Marc M. 2013, Rethinking the Vulnerability of Minority Populations in Research, *Public Health Ethics*, 103(12): 2141–2146.

Rogers, Wendy, Mackenzie, Catriona, Dodds, Susan 2012, Why Bioethics Needs a Concept of Vulnerability, *International Journal of Feminist Approaches to Bioethics*, 5(2): 11–38.

Schonfeld, Toby 2013, The Perils of Protection: Vulnerability and Women in Clinical Research, *Theoretical Medicine and Bioethics*, 34: 189–206.

Schrems, Berta 2014, Informed Consent, Vulnerability and the Risks of Group-Specific Attribution, *Nursing Ethics*, 21(7): 829–843.

Siriwardhana, Chesmal 2015, Rethinking Vulnerability and Research: Defining the Need for a Post-Research Ethics Audit, *Asian Bioethics Review*, 7(2): 188–200.

Straehle, Christine 2016, Vulnerability, Health Agency and Capability to Health, *Bioethics*, 30(1): 34–40.

Tavaglione, Nicola, Martin, Angela K., Mezger, Nathalie, Durieux-Paillard, Sophie, François, Anne, Jackson, Ives, Hurst, Samia A. 2015, Fleshing Out Vulnerability, *Bioethics*, 29(2): 98–107.

ten Have, Henk 2014, Vulnerability as the Antidote to Neoliberalism in Bioethics, *Revista Red-bioética*, 5,1(9): 87–92.

ten Have, Henk 2015, Respect for Human Vulnerability: The Emergence of a New Principle in Bioethics, *Journal of Bioethical Inquiry*, 12(3): 395–408.

ten Have, Henk 2016, *Vulnerability. Challenging Bioethics*, New York: Routledge.

Thompson, Clay, Bacon, Alison M., Auburn, Timothy 2015, Disabled or Differently-Enabled? Dyslexic Identities in Online Forum Postings, *Disability and Society*, 30(9): 1328–1344.

Toader, Elena, Damir, Dana, Toader, Tudorel 2013, Vulnerabilities in the Medical Care, *Procedia – Social and Behavioral Sciences*, 92: 936–940.

Turner, Bryan S., Dumas, Alex 2013, Vulnerability, Diversity and Scarcity: On Universal Rights, *Medicine, Health Care and Philosophy*, 16: 663–670.

UNESCO 2005, *Universal Declaration on Bioethics and Human Rights*, Paris.

UNESCO 2008, *Report of the International Bioethics Committee of UNESCO (IBC) on Consent*, Paris.

UNESCO 2010a, *Meeting of the International Bioethics Committee's Working Group on Human Vulnerability and Personal Integrity*, Paris: SHS/EST/10/CIB/WG-1.

UNESCO 2010b, *Preliminary Draft Report on the Principle of Respect for Human Vulnerability and Personal Integrity*, Paris: SHS/EST/CIB-17/10/CONF.501/2.

UNESCO 2011, *Draft Report of IBC on the Principle of Respect for Human Vulnerability and Personal Integrity*, Paris: SHS/EST/CIB-17/10/CONF.501/2 Rev.

UNESCO 2013, *The Principle of Respect for Human Vulnerability and Personal Integrity: Report of the International Bioethics Committee of UNESCO (IBC)*, Paris.

UNESCO, 2014, Report of the IBC on the Principle of Non-discrimination and Non-stigmatization, Paris.

UNESCO, 2015, *Report of the IBC on the Principle of the Sharing of Benefits*, Paris.

UNESCO 2016, *Educational Resources*, <www.unesco.org/new/en/social-and-human-sciences/themes/bioethics/ethics-education-programme/activities/educational-resources/> accessed 22-03-2016.

Wendell, Susan 1996, *The Rejected Body*, New York: Routledge.

World Medical Association 2008, *Declaration of Helsinki – Ethical Principles for Medical Research Involving Human Subjects*, Geneva.

World Medical Association 2013, *Declaration of Helsinki – Ethical Principles for Medical Research Involving Human Subjects*, Geneva.

Index

Printed and bound by CPI Group (UK) Ltd, Croydon, CR0 4YY

01/11/2024

01782600-0007